SO-EGI-260

TIME AND AGAIN

TIME AND AGAIN

A Systematic Analysis of
the Foundations of Physics

Marinus Dirk Stafleu

Wedge Publishing Foundation,
Toronto, Canada

Sacum Beperk
Bloemfontein

© Copyright 1980: M. D. Stafleu
Couwenhoven 65-31, Zeist, Netherlands
All rights reserved
Set in 10 on 11 pt. Times Roman
Printed and bound by
National Book Printers
Goodwood, Cape
First impression 1980

ISBN 0 949994 55 3

ISBN 0 88906 108 4 (Studiereeks)

Prof. D.F.M. Strauss
INLEIDING TOT DIE KOSMOLOGIE
Wetenskaplike Studiereeks Nommer 1

M.D. Stafleu
TIME AND AGAIN
Wetenskaplike Studiereeks Nommer 2

(gesamentlike uitgawe met Wedge
Publishing Foundation, Toronto, Canada)

Preface

Time and Again is intended to prove that physics, as far as its foundations are concerned, is little more than time keeping. Time is understood as a lawful pattern of relations between things and events. It is this pattern which forms the subject matter of this study.

Time and again, philosophers of science have been in search of a unifying principle in the foundations of physics. Time is not such a unifying principle. We want to show that time is rather a diversifying principle. We do not wish to find unity, but to account for the diversity of nature.

Time and again we shall argue that the relations of time are not based on some conventions, and that the laws of physics are not merely convenient patterns of thought. We shall provide arguments for the view that physics is an historical endeavour, intending to open up the creation by discovering laws and applying them to physical reality.

Time and again philosophers have tried to present the foundations of science on an a priori basis. We want to discover them by a close scrutiny of the physical sciences and their history. Only one thing will be taken for granted: the lawfulness of the creation. Different philosophical schools can be distinguished by the way they give account of this lawfulness.

We want to study the foundations of physics in a systematic way. Hence we begin exposing a hypothetical framework for our investigation. It is derived from the *Philosophy of the Cosmonomic Idea*, based on Christian principles. Yet it has an empirical character, apart from its view on the Origin of the creation and its lawfulness.

Time and again I have benefitted from many people who have contributed to this work. I shall not mention all my teachers, colleagues and friends, but only them who have directly helped to shape this book. Parts of its manuscript were critically read by Mr C. C. Adams, Dr W. J. Alberda, Dr J. C. Bellum, Dr J. Cook, Mr C. Jongsma, Dr L. Jonker, Mr D. Judd, Mr P. Mahaffy, Dr K. Piers, Mr A. Tol, Dr R. E. VanderVennen, and Dr J. Wong. Most thanks are due to Dr A. Leegwater, who revised the manuscript entirely, and to Dr R. E. VanderVennen, who continuously encouraged me to publish it. Finally, I am greatly indebted to my wife and my children, who have had to endure many hours of absentmindedness on my part.

Note: Author's names in the footnotes refer to the Name Index and Bibliography at the end of the book. If necessary, different publications by the same author are distinguished by a capital (A, B, C . . .).

Contents

1. Framework

1.1 *Foundations research*

Although there are many handbooks and textbooks of physics as well as numerous monographs and papers on special topics, until recently, there have been few books dealing with the structure of physics in a manner which goes beyond a mere commentary on its methods. Many of the texts, of course, are extremely important for the understanding of physics, and are fascinating because of new vistas explored or admirable because of clarity in expounding older views. However, there remains surprisingly few investigations into the basic structure and coherence of the physical sciences.

There is, perhaps, a historical explanation for this. The old philosophy of nature ("Naturphilosophie") assumed that the foundations of physics could be derived from "immediately evident truths" of an a priori, transcendent and necessary character. It was thought that these truths could be understood without the need of experimental verification. Especially in 19th-century central-European circles, attempts were made to build a structure of physics on such speculative foundations. Hegel is notorious in this respect. In time, however, it became clear that many of these "self-evident truths" were in fact false. In reaction, many late 19th- and 20th-century physicists rejected outright any a priori philosophical bias for their work – and willy-nilly became adherents of another philosophy: some variant of positivism (neo-positivism, Vienna school, analytical philosophy, instrumentalism, operationalism, conventionalism). Assuming that the *content* of science is "positive fact" which must be taken for granted, whereas the *structure* of science is determined by its *methods*, most positivist philosophers are interested only in the latter.[1] Hence for positivists, philosophy of science is not a matter of ontology or epistemology, but rather a matter of methodology.

The study of the foundations of physics has traditionally been called "metaphysics", but, since the beginning of the 19th century,

---

1. On positivism, see Frank; Kolakowski; Von Mises; Popper F.

this term has become discredited because of its speculative implications. Currently, this kind of study is usually referred to as "foundations research". Bunge defines its goal as being twofold: "To perform a critical analysis of the existing theoretical foundations (of physics), and to reconstruct them in a more explicit and cogent way".[2] The critical analysis has three tasks:

"(a) To examine the philosophical presuppositions of physics;
(b) To discuss the status of key concepts, formulas, and procedures of physics;
(c) To shrink or even to eliminate vagueness, inconsistency, and other blemishes."[3]

Similarly, the task of reconstruction, according to Bunge, has three aspects:

"(a) To bring order to various fields of physics by axiomatizing their cores;
(b) To examine the various proposed axiomatic foundations;
(c) To discover the relations among the various physical theories."[4]

For Bunge, the most important tool of foundations research is axiomatization. In this context, ". . . 'axiom' means *initial assumption* not self-evident pronouncement. There need be nothing intuitive and there is nothing final in an axiom . . ."[5] Axiomatization of physical theories ". . . does nothing but organize and complete what has been a more or less disorderly and incomplete body of knowledge: it exhibits the structure of the theory and makes its meaning more precise."[6]

However, since axiomatization is more an investigation of theories than of physics, it is unlikely that foundations research can be exhausted by formulating axioms. In the first place, axiomatization can only be applied to partial theories[7] such as classical mechanics, classical electromagnetism, thermodynamics, special and general relativity – to mention fields in which this type of foundations research has been carried out more or less successfully.[8] Moreover, as Whiteman observes, ". . . a conceptual system such as Euclidean

2. Bunge B 1.
3. Bunge B 1, 2.
4. Bunge B 2.
5. Bunge B 64; Bunge E.
6. Bunge B 68, 69.
7. See Segal, in: Henkin 341: ". . . no axiom system is secure if it does not treat a closed system."
8. Bunge H; Noll.

geometry may be subjected to innumerable axiomatizations, all hazy in different ways."[9] In this book we shall not be primarily interested in partial theories, and we shall make use only occasionally of available axiomatizations. Our main focus will be with an ordering scheme of all aspects of the physical sciences – i.e., with the third of Bunge's "constructive tasks" of foundations research. It is very doubtful whether such an ordering scheme could be axiomatized in any sense, since any axiomatization would itself probably depend on such a scheme, whether explicitly recognized, or implicitly assumed. In our discussion, the partial theories neither are placed alongside one another, nor will they be deductively subsumed. They turn out to be interdependent and it is especially this mutual dependence which will be our subject matter.

A second reason for rejecting axiomatization as the main tool of foundations research is this: any modern axiomatization system familiar to me relies heavily on set theory, as well as on a formal logic making use of set-theoretic methods. This appears to betray a strong influence of Aristotelian philosophy of science, according to which science means the designation of classes and their mutual relations. This Aristotelian influence may be spurious; nevertheless, the approach relies heavily on logic, the laws of which are supposed to be true (if only "vacuously true") and a priori valid tools in foundations research. We shall consider set theory to be a mathematical theory, and insofar as logic makes use of it, logic has a retrocipatory character (cf. Sec. 2.5). Thus, sets and classes as mathematical entities should find a place within the general ordering scheme we are seeking. This implies that we cannot accept set theory and its dependent, axiomatization as the basis of our research into the foundations of physics, though both will play an important role in this book.[10]

From the above quotations it should be clear that Bunge's extreme emphasis on logical methods does not imply a purely deductive approach to physics, for his axioms must be found in existing physical theories. Still, he seems to adhere to the medieval idea that everything special is contained in the general. Cantore's "inductive-genetic" approach presents a somewhat different view: "First, the approach should be *inductive* . . . the philosophical approach to science, to be successful, should concentrate on the detailed study of individual, fully developed theories. Secondly, the approach has to be *genetic*.

---

9. Whiteman 104, 105; cp. Bunge B 66; Jammer, C, Ch. 9, especially p. 120.
10. Cp. Dooyeweerd C 59. Gödel's theorem concerning the consistency and the completeness of axiomatized theories also shows some limitations of this method; cf. Gödel; Bunge B 64.

Each scientific theory arises out of a slowly growing body of information. Hence the nature of the scientific endeavour and its achievements cannot be properly realized unless one follows the developments of individual theories as they gradually unfold and develop in time."[11] This points out my third objection to Bunge's position: the historical development of a theory must also be accounted for in foundations research.

Finally, I wish to direct a few comments to Bunge's first "critical task" of foundations research – viz., to examine the philosophical presuppositions of physics. First, it must be emphasized that there exists no unique set of philosophical presuppositions. Second, no examination of such presuppositions can itself be philosophically neutral. Bunge himself seems to be more clear about the philosophies which he rejects (positivism, operationalism) than he is about his own philosophical position (realistic objectivism, or critical realism[12]). This vagueness about one's own philosophy is not unusual among workers in foundations research. Since the beginning of this century it has become abundantly clear that mathematics and physics, and more specifically, investigations into their foundations, are not free from philosophical assumptions, which, in turn, depend on one's world and life view. Recognition of this has led to a more or less peaceful coexistence of different philosophical traditions in mathematics (logicism, formalism, and intuitionism)[13] and in physics ((neo-)positivism, operationalism, realism, conventionalism, materialism, and phenomenalism).[14] Since this book is not concerned primarily with philosophy, a complete criticism of any of these philosophical systems is impossible, but, at times, we shall have occasion to confront our view with the views of others.

The philosophical position from which the book is written is the so-called *Philosophy of the Cosmonomic Idea*, developed by Professors H. Dooyeweerd and D. H. Th. Vollenhoven at Amsterdam, during the second quarter of this century.[15] In contrast to philosophical fashion, this philosophy does not degenerate into a kind of methodology. Growing out of the reformed biblical "ground motive" of creation, fall into sin, and redemption through Jesus Christ, it is a

11. Cantore 5.
12. Bunge B 44, 49, 58, 287; Bunge E.
13. Fraenkel, Bar Hillel.
14. These are contemporary philosophies. For an enumeration of eight mostly historical views on the relation of natural philosophy and science, see Beth C, Ch. 3. See also Losee A.
15. Dooyeweerd B, C, D, E; for an introduction to Dooyeweerd's philosophy, see Kalsbeek.

rather complicated attempt to account for the full complexity of created reality. Not only is this philosophy a systematic investigation into the structure of created reality and our knowledge thereof, but it also tries to account for the temporal development of created reality. For readers of this book it would be helpful to have prior knowledge of this philosophy. However, since only part of its elaborate system is needed for our analysis of the structure of physics, and since, insofar as we need it, we will elaborate the system in the course of this book, such prior knowledge is not strictly necessary. In this introductory chapter an outline of the general framework within which our discussion takes place will be given. We do not wish to present this philosophy as an a priori truth; on the contrary, to a large extent, its applicability must be demonstrated by studies such as the one undertaken in this book. Hence we invite the reader to understand this introductory chapter as a provisional outline of a working hypothesis which is to be tested in the following chapters. In order to understand the structure of the physical sciences, we are in need of a philosophical system which makes possible *a systematic analysis of the foundations of physics,* including its history.

1.2 *Three basic distinctions*
Three central, recurring themes can be recognized in the history of scientific philosophy: the search for *truth,* the search for *order,* and the search for *structure.* The first is mainly a philosophical concern, and deals with the relation of laws and which is subjected to them, the status of law (the nominalism-realism controversy), the possibility of human knowledge, and the methodology of science. Its central problem is to account for the lawfulness of creation. The search for *order* and *structure* forms the core of science, and here one deals with questions such as: *Are there general modes of experience which provide an order for everything within the creation,* and if so, *which are these universal orders of relation? How can stable things exist, and how can they change?* The question of structure already surfaced in Greek philosophy and is still prominent in modern physics and biology, whereas the problem of order did not appear until post-Renaissance science. These three themes, though they cannot be treated separately, are irreducible to each other, and they lead to the introduction of three basic distinctions which form the skeleton of our philosophical theory.

(*a*) *The distinction of law and subject* (Sec. 1.3) is basic to all sciences, though it is not always explicitly recognized as such. Every science worth its name is concerned with laws. These laws are concerned either with more or less concrete things, events, signs, living beings,

5

artifacts, social communities, etc., or with more or less abstract concepts, ideas, constructs, etc. These things which are subjected to law are commonly referred to as "objects", but, for reasons to be explained in Sec. 1.6, we shall refer to them as "subjects" – i.e., beings subjected to laws.[16] Though we will not deal with this problem extensively, we shall make a few comments about it, occasionally.

(b) *The distinction of typicality and modality* (Sec. 1.4). As mentioned above, we distinguish those "subjects" which are more or less concrete from those which are more or less abstract. This distinction is mirrored in the one between *typical*, special laws, which apply to a limited class of subjects, and *modal*, general laws, which hold for subjects of a more abstract character. Our first distinction (law and subject) is frequently identified with the distinction of universals and individuals. However, this identification is inadequate and too crude, since the distinction of typical and modal laws also implies a universal-individual duality. For the same reason, laws cannot be identified with classes or sets,[17] although special laws define classes. Modal laws, however, do not, and therefore cannot be found by generalization: they must be inferred by abstraction.

(c) *The various modal aspects* (Sec. 1.5). Various solutions to the problem of the general modes of experience have been presented. However, most of these attempt to solve the problem in terms of a *single* principle of explanation. This has led to a proliferation of "isms" in philosophy and science: arithmeticism (Pythagorean tradition), geometricism (Descartes' "more geometrico"), mechanism (Huygens, Leibniz, Maxwell), evolutionism, vitalism, behaviorism, logicism, intuitionism, historism, etc. In contrast to this trend, we shall attempt a solution in terms of *several* mutually irreducible modes of experience. Dooyeweerd was the first to recognize that the modal laws can be grouped into several "law spheres" or "modal aspects". Each modal aspect is of a general, universal character, but is irreducible to any other. We shall concern ourselves primarily with only four modal aspects which we designate as the numerical, the spatial, the kinematic, and the physical aspect.

16. We shall usually use the term "subject" as a noun. Instead of the adjective "subject" we use the somewhat uncommon "subjected". E.g. we write that a subject is *subjected* to law, not is *subject* to law. Cf. Dooyeweerd B 108.

17. Moreover, classes and sets have both a law side and a subject side, and cannot be identified with either one.

6

The three basic distinctions which we have made above are neither dependent on each other nor reducible to each other. We may picture them as being mutually orthogonal, like the three axes in a Cartesian coordinate system. The three distinctions, though independent and irreducible, must be studied simultaneously, since they interpenetrate one another. It is not possible to discuss one of them without taking into account the other two. In the following sections I shall discuss these distinctions more extensively. During the discussion I shall point out several distinct *"aims of science"* which differ from one another to the extent that different viewpoints are possible within our systematics. I shall argue that each distinction implies a twofold direction (which I have tried to indicate by arrows in the diagram below). In the "horizontal plane" (orthogonal to the law-subject coordinate) these directions refer to the historical development of science (Sec. 1.7). This means that the systematics to be discussed below has a dynamic, not a static character.

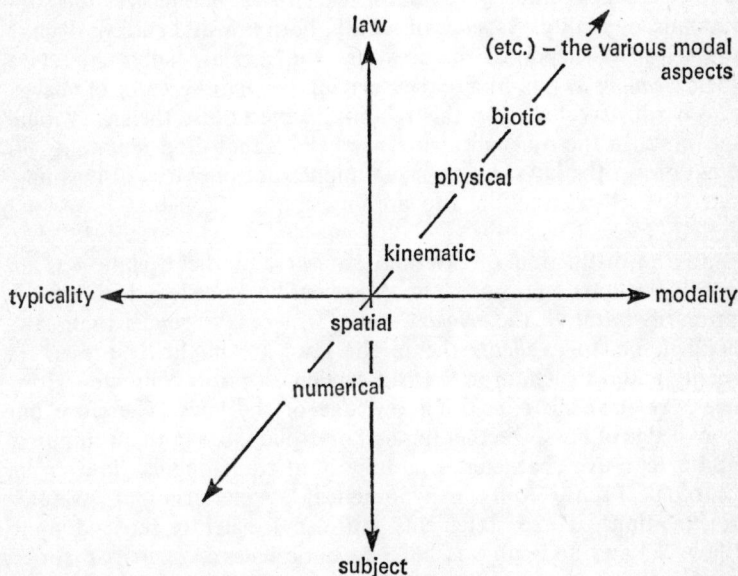

1.3 *Law and subject*
The first basic distinction in our investigation of created reality is that of its law side and its subject side. In every philosophy, rightly called scientific, this distinction is explicitly or implicitly made; without it, science is impossible. This distinction is not merely a scientific, epistemological one, it is rooted in the creation itself. In fact, not only science, but all our life would be impossible without the

7

awareness (mostly subconscious) of laws, distinct from subjects. In our lifetime we encounter things, animals, plants, men, human societies, organizations, and, above all, ourselves. All of these I refer to as "subjects", i.e., they are all subjected to some kind of law. Indeed, it is precisely because of these structural laws that we can distinguish the various subjects from one another, and explain and predict their behaviour. Without having some intuitive idea of structural laws which hold for plants, animals, etc., it would not be possible even to speak of them. We would be unable to perform even the simples acts of life if we had no idea of, and confidence in, lawfulness.

There are no subjects without structural laws. Every subject is constituted by some law, and is related to other subjects by laws. The reverse is also the case. There are no laws without subjects (either possible or actual). The function of a law is to be valid for some or all subjects. These two sides of reality are correlative. As a result, we must avoid both rationalism, which overrates the law side of creation, and irrationalism, which over-emphasizes the subject side of reality. As *sides* of reality, both law and subject display the self-insufficiency of the creation. Via the law, subjects receive their *meaning* by pointing to their origin, the Divine creator of heaven and earth. Isolated from this relation, subjects lose their creational meaning. In the other direction, we believe that God maintains his creation via the laws. Lawlessness implies not only loss of meaning, but also self-destruction into nothingness.

Both the distinction of law and subject and their relation is an ontological matter. Since our pre-scientific knowledge of laws is primarily intuitive, the *primary aim of science* is to render these laws explicit, i.e. to *explicate* them. The laws are implicitly present in reality and, as such, are of a transcendent, a priori character. However, we have no a priori knowledge of the laws. Therefore our knowledge of laws, whether implicit or explicit, has both an empirical and a tentative character. It is important to distinguish laws in an ontological sense from our hypothetical law statements in scientific formulations. These statements are also frequently referred to as "laws". Laws and subjects have an ontic nature, whereas theories, models and facts have an epistemic nature.[18] Thus, while electrons

18. Bunge A 248ff; B 44; Bunge A 249 defines laws as ". . . the immanent patterns of being and becoming . . .", law statements as ". . . the conceptual reconstructions . . ." of laws. The relation of law and subjects, and the status of theories, models, facts, induction, deduction, and reduction are objects for epistemological research. See e.g. Hempel A, B, C; Nagel B; Popper A, B; Stegmüller.

have always existed and the laws concerning them have always been valid, an electronic *theory* and the *fact* of the existence of electrons did not appear until 1896. Prior to that year the existence of an electron was not a fact. Theories, hypotheses, models and facts, though bound to the creation order, are human inventions. But laws and sub-human subjects exist independently of human knowledge.

We can discover laws, e.g., by induction, because they are related to subjects and we test the validity of our law statements by confirmation with facts. This state of affairs, however, does not mean that we can reduce laws to subjectivity. This was most clearly recognized by Hume, who argued that the "inductive assumption" concerning the possibility of finding laws by induction cannot be justified by experience. Hume insisted that there is no epistemic proof that laws concerning future events can be inferred from regularities observed in the past.[19] This discovery resulted, on the one hand, in a sceptical attitude concerning the very possiblity of science and, on the other, in the conviction that laws have a merely epistemic status. We share neither of these views. For us, the possibility of discovering laws is based on faith in the lawfulness of reality and in a God who faithfully maintains his laws. Admittedly, the lawfulness of reality cannot be proved. It is an *a priori* of all human experience, including scientific experience.[20] According to the Philosophy of the Cosmonomic Idea, different philosophical systems can be characterized according to their respective views on the status of law.[21] Thus the *name* of this philosophy does not lay a claim on the "cosmonomic idea". Rather it pleads for the recognition that any philosophical system must account for the lawfulness of reality. Such an account does not have a scientific starting point, but a religious starting point, which has scientific consequences.

The positivist view that the truth of law statements can be established by verification of their factual consequences has been criticized by Popper.[22] However, Popper's "falsifiability criterion", though a correction to the positivist view, is only sufficient to "demarcate"

19. Hume; Braithwaite, Ch. 9; Harris B 39, 40; Kolakowski 42-59; Losee A 101-106; Popper E 1-31, 85-105; Russell C 634-647. Dooyeweerd B 275ff observes that Hume's scepticism has a methodological significance, intended to reinforce his psychological ideal of science.

20. Even the existence of subjects outside ourselves cannot be proved, as was shown by solipsism; cf. Russell B 27ff.

21. See Dooyeweerd B 93ff. For a discussion of the status of Newton's second law of motion (which could serve as an illustration of this assertion), cf. Hanson B, Ch. 5.

22. Popper A, Ch. 1.

scientific from non-scientific law statements. Regardless of how much evidence may "corroborate" a natural law statement, acceptance of the statement as law is always a matter of faith. A law statement is ultimately *believed* to be true, because of convincing evidence supporting it. This belief does not *prove* that the law statement is true, for such proof does not exist. This belief is neither individual nor irrational; it is *communal*, i.e., the community of scientists decides on the faithfulness of the empirical evidence and the acceptibility of physical theories.[23] To perform this task the scientific community organizes societies, journals, etc., in which the evidence is judged and debated, according to unwritten codes. Even then, the truth of any law statement cannot be absolutely proved or disproved. Indeed, in many cases, scientific research is initiated because someone (on quite rational grounds) does not believe the accepted views on some particular subject.

Ultimately, an acceptance of the truth of law statements and empirical evidence is based on belief, both in the reliability of our colleagues, and in the lawfulness of the creation. Hence there is room for the rejection of formerly held law statements, and the critical reconsideration of older evidence in the light of new evidence or insights. This same state of affairs applies to our consideration of subjects. Because of the correlation of law and subject, our knowledge of facts is always theory-laden. Thus, at present, it seems quite certain that electrons and stars are real existing entities, whereas we are far less sure about the existence of quarks and quasars. In these cases, too, the degree of certainty depends upon the availability of independent, reliable evidence. In our opinion, both laws and subjects are "discovered", implying an active role for the scientific explorer. How theories are found or invented is not well understood. The scientist's phantasy and genius itself is subjected to historical and psychological research, and certainly cannot be reduced to simple logical rules for deduction and induction.[24]

Though the number of laws may be infinite, they are not all independent, and it is often possible to deduce one law statement from others. In this case we say that the former is reduced to the latter. The reduction of laws and, conversely, the deduction of new laws and their consequences for subjects is the *second aim of science.*

23. The influence of communal belief on accepted theories has been emphasized by Kuhn B, D; see also Bunge B 70; Harris B; Ziman A, B.

24. Kuhn B, Ch. 2; Feyerabend E; Lakatos A; Holton A, Ch. 10; Holton B, Ch. 3; Finocchiaro.

Axiomatization can be a very helpful tool in investigating the possibility of such reduction schemes. Attempts to reduce all laws to a single principle have been made in every epoch of philosophy, beginning with Thales' "Everything is made of water".[25] In classical physics an attempt was made to explain all physical phenomena from the motion of unchangeable pieces of matter. We share none of these reductionist views – in fact, we shall assume that the mutual irreducibility of the various modal aspects precludes its possibility.

1.4 *Typicality and modality*
In addition to the distinction of law and subject, it is very fruitful to introduce a second basic distinction, that of typicality and modality. Hereby, we distinguish laws which are valid for a limited class of subjects (typical laws) from those which are valid for all kinds of subjects (modal laws). Typical laws, in principle, delineate the class of subjects to which they apply, describing their structures and typical properties. Examples of such laws are the Coulomb law (applicable only to charged subjects), the Pauli principle (applicable only to fermions), etc. Often the law describing the structure of a particular subject (e.g., the copper atom) can be reduced to some more general typical laws (e.g., the electromagnetic laws in quantum physics). On the other hand, modal laws are those which have a universal validity. For example, the law of gravitation applies to all physical subjects, regardless of their typical structure. We call them *modal* laws because, rather than circumscribing a certain class of subjects, they describe a *mode* of being, relatedness, experience, or explanation.

This distinction is also relevant to the way in which different laws are discovered and formulated. Whereas typical laws can usually be found by induction and generalization of empirical facts or lower-level law statements, modal laws are found by abstraction. Euclidean geometry, Galileo's discovery of the laws of motion and the subsequent development of classical mechanics, and thermodynamic laws are all examples of laws found by abstraction. This state of affairs is reflected in the use of the term "rational mechanics", in distinction from experimental physics.

At first sight the distinction between typicality and modality appears to apply only to laws. Indeed, all concretely existing things, events, organisms, etc., have some typical structure. However, even as modal laws are found by abstraction, *modal subjects*, which are abstracted from any typical and individual properties, are also found to exist. The modal subjects (so-called because they are ex-

25. Russell C 44, 61.

clusively subjected to modal laws) are indispensable in science for the ordering of our experience. Numbers, spatial figures, inertial systems, wave packets, and isolated systems are all examples of modal subjects. They do not "exist" in any *concrete* sense, since they lack any individuality and typicality. Nevertheless, in the sense of belonging to created reality, these subjects are perfectly real – they are abstract modal subjects. In our scientific work we must abstract from concrete, individually existing things, events, etc. We must disentangle typical laws in order to discover modal laws. This process could not be carried out without the use of abstract modal subjects. We may call abstraction the *third aim of science*, which includes the formulation of modal, universal laws, as well as the modal analysis of concrete reality on both the law side and the subject side.[26]

The distinction of typicality and modality is, however, not merely an epistemological one, for though there is a plurality of laws and subjects, there is only one reality. This means that even though subjects may have widely differing typical structures, they must be related in a general way. It is these general (thus modal) subject-subject relations which come to the fore when we study modal laws (cf. Sec. 1.5). For instance, two physical subjects, regardless of their typical, individual structures, are always related, since they must have a certain spatial distance and a certain relative velocity. But in order to investigate these general relationships, we must deprive the subjects of their typicality – i.e., modal laws have correspondent modal subjects.

The distinction of typicality and modality is not of great importance in mathematics because, as we shall see, this science is concerned primarily with purely modal laws. As a result, many philosophers and mathematicians have assumed that mathematics is just a branch of logic. However, physics is concerned not only with modal laws, but also with typical laws regarding the typical structure of atoms, crystals, stars, etc. Therefore, in the physical sciences, more so than in mathematics, the distinction of general, modal laws and specific, typical laws becomes important. This does not mean that mathematics is less empirical than physics. It would be an oversimplification to state that mathematics, as a "formal science", is concerned with modal, a-typical laws, whereas physics, as an

26. We relate "abstraction" to the distinction of modality and typicality. As we shall see, one may also speak of abstraction in another sense if one studies a modal aspect without considering the succeeding modal aspects. In experimental physics one *isolates* a system (and in theories one isolates a problem) – in order to keep it under control – by keeping some influences constant while changing one or more others.

"empirical science", deals with typical laws and, therefore, with concrete reality.[27]

A *fourth aim of science* is the reconstruction or synthesis of typical laws. After finding some general characteristics of a certain structure (e.g., the number of protons and neutrons in a helium nucleus) by modal analysis, an effort to find the typical law for this system may be made. Since modal laws are too universal to form any typical structure, the starting point for the reconstruction cannot be taken solely in the modal laws themselves. As it happens, in physics, in addition to purely modal gravitational interaction one must also consider electromagnetic interaction and two types of nuclear interaction. Despite many efforts toward the development of a "unified field theory", these fundamental interactions cannot be reduced to one another. With the help of modal laws and these typical interactions of a sufficiently general character, an enormous number of typical structures may be formed: nuclei, atoms, molecules, crystals, particles, quasi-particles, etc. Investigations of these structures reveal both sides of the modality-typicality distinction: abstraction and reconstruction, analysis and synthesis. Without the existence of the irreducible fundamental typical interactions, typical laws could be subsumed under modal laws. Because of their irreducibility, the distinction of typicality and modality must be recognized as being "orthogonal" to the distinction between law and subject.

Our distinction of typicality and modality appears in several other philosophical systems in one form or another. Campbell distinguishes typical laws from other laws. He calls typical laws ". . . laws of the kind which assert the properties of a kind of system . . . The 'classificatory' sciences differ from other sciences in that they confine themselves to laws of this type . . ."[28] Margenau[29] speaks of the "immediately given" from which a scientist passes to "orderly knowledge" by the formation of "constructs". Between the former and the latter there are "rules of correspondence" and there is a "circuit of empirical confirmation". Bunge[30] states "Every physical idea is expressed in some language and has a logical structure and a context of meaning."[31] The language has a (modal) syntax or grammar and, via

27. Cp. Bunge B 28. It is also useless to consider mathematics as "analytical" and physics as "synthetic" (cp. Jauch 70).
28. Campbell A 56, 57.
29. Margenau A, Ch. 3-6.
30. Bunge B, Ch. 1.
31. Bunge B 9; see also Jammer E 10ff.

a semantics, is connected with reality. From our point of view, we may recognize this because the logical and lingual modal aspects have a universal character, but we avoid the pitfall of absolutizing them. There are other universal and irreducible modal aspects as well. While recognizing that both logic and language are indispensable for the human act of scientific abstraction and reasoning, we shall not concern ourselves further with these epistemological problems.

1.5 *The modal aspects*

The theory of the modal aspects is one of the most important chapters in the Philosophy of the Cosmonomic Idea developed by Prof. H. Dooyeweerd. He says:

". . . our theoretical thought is bound to the temporal horizon of human experience and moves within this horizon. Within the temporal order, this experience displays a great diversity of fundamental modal aspects, or modalities which in the first place are aspects of time itself. These aspects do not, as such, refer to a concrete *what*, i.e., to concrete things or events, but only to the *how*, i.e., the particular and fundamental mode, or manner, in which we experience them. Therefore we speak of the modal aspects of this experience to underline that they are only the fundamental *modes* of the latter. They should not be identified with the concrete phenomena of empirical reality, which function, in principle, in all of these aspects. Which, then, are these fundamental modes of our experience? I shall enumerate them briefly.

"Our temporal empirical horizon has a numerical aspect, a spatial aspect, an aspect of extensive movement, an aspect of energy in which we experience the physico-chemical relations of empirical reality, a biotic aspect, or that of organic life, an aspect of feeling and sensation, a logical aspect, i.e., the analytical manner of distinction in our temporal experience which lies at the foundation of all our concepts and logical judgments. Then there is a historical aspect in which we experience the cultural manner of development of our societal life. This is followed by the aspect of symbolical signification, lying at the foundation of all empirical linguistic phenomena. Furthermore there is the aspect of social intercourse, with its rules of courtesy, politeness, good breeding, fashion, and so forth. This experiential mode is followed by the economic, aesthetic, juridical and moral aspects, and, finally, by the aspect of faith or belief."[32]

Dooyeweerd argues that the several modal aspects are *mutually*

---

32. Dooyeweerd E 6, 7; see also Dooyeweerd B 3; on the criterion of a modal aspect, see Dooyeweerd C, Ch. 1.

irreducible. Because of the genetic character of scientific knowledge, the designation of the various modal aspects must always be of a tentative and hypothetical character. Dooyeweerd himself did not distinguish the kinematic from the physical modal aspect until 1953.[33] The distinction of two mutually irreducible modal aspects is based on an analysis of our contemporary knowledge. We shall report on such an analysis for the first four modal aspects, which, for the sake of brevity, we designate as the numerical, the spatial, the kinematic, and the physical modalities.

In science, the different modes of experience can be different *modes of explanation* as well. As a result, we are provided with a possible distinction of the special sciences on an ontological basis, at least insofar as a special science can be characterized by one of the irreducible modes of explanation. In principle, each modal aspect has a corresponding special science: arithmetic or algebra with the numerical aspect, geometry with the spatial aspect, kinematics with the kinematic aspect, physics (including chemistry and astronomy) with the physical aspect, biology with the biotic aspect, etc.[34] This classification is not exhaustive, however, since some sciences (geology, for example), study certain structures from the viewpoint of several modal aspects, no single one of which takes a leading role.

Temporal reality is a multiply-connected pattern of relations. Although many of these relations have a typical structure, it is only possible to understand the unity, i.e., the mutual relatedness of all subjects in the creation, if at least some of these relations are of a modal, universal character. All concrete existing things, events, etc., have mutual numerical, spatial, kinematic, and physical relations, and it is these mutual relations that make it possible for us to become aware of and understand these subjects. We are, therefore, entitled to speak of the modal aspects as *universal modes of temporal relations.*[35]

Within this modal relatedness we may distinguish a law side and a subject side. On the law side, in each modal aspect, we find a distinct modal order, which is correlated with a modal subject-

33. Dooyeweerd C 98; Dooyeweerd's view on these two modal aspects is not shared by all adherents of his philosophy. There is also some disagreement on several other modal aspects.

34. The prevailing positivist view reverses the creational order by stating that the sciences must be classified according to their methods (cf. Margenau A 46).

35. The theory of time discussed in this book differs slightly from Dooyeweerd's; cp. Dooyeweerd A, B 22-34.

subject relation on the subject side. In the *numerical* aspect the modal order is the serial order of smaller and larger, or earlier and later. This modal order originally correlates with the numerical difference or ratio of two numbers, as modal, numerical subject-subject relations. The modal order in the *spatial* modal aspect is that of simultaneous coexistence, which is correlated with the relative spatial position of two subjects on the subject side. In the *kinematic* modal aspect the modal order of time flow is correlated with subjective relative motion, and in the *physical* aspect the modal order appears as irreversibility, which is correlated with the physical interaction of two or more subjects on the subject side.

As it turns out, the modal order in every modal aspect refers to our common understanding of *time*, since earlier or later, simultaneity, time-flow, and irreversibility are all acknowledged temporal relations. At first sight, the same cannot be said of the modal subject-subject relations such as relative position and interaction. However, we shall see that on the subject side, the opened-up numerical subject-subject relations (anticipating other subject-subject relations) most closely approximate what we usually refer to as "time". This is most clearly shown by an analysis of the historical development of time measurement, at least insofar as such a development can be reconstructed. Initially, time measurement was simply done by *counting* (days, months, years, etc.). Later, time was measured by the relative *position* of the sun or the stars in the sky, with or without the help of instruments such as the sundial. In still more advanced cultures, time was measured by utilizing the regular *motion* of more or less complicated clockworks. Finally, in most recent developments time is measured via *irreversible* processes, for example, in atomic clocks.

In a scientific context, however, it is inadequate to work with either a simple common notion of time, or a merely objective representation of subjective relations. All modal subject-subject relations as well as the modal orders to which they are subjected must be recognized as being "temporal". *Time relates all subjects to each other under a universal law of order.* The question as to whether time is relational or absolute in some sense has long been debated and still has not been settled.[36] We reject any notion of absolute time. "Absolute time" infers a unique universal reference system. We shall show that our theory of modal time requires the existence of several classes of reference systems – none of them unique, all of them universal – which allow an objective description of our world.

36. For an anthology on the problem of time, cf. Gale; Zeman.

Although the modal aspects are mutually irreducible, they are neither unconnected nor independent. The modal aspects display a serial order. As a result we can speak of "earlier" and "later" modal aspects in the sense that a later modal aspect presupposes the earlier ones. For example, the spatial modal aspect presupposes the numerical aspect. If this were not so, it would not be possible to speak of *three*-dimensional space, the *four* sides of a square, or any other numerical attribute of spatial functioning. In a similar way, the spatial aspect is presupposed by the kinematic modal aspect, which in its turn, is presupposed by the physical aspect. Similarly, the biotic aspect presupposes the physical aspect, and so forth. Terminologically, we say that the later aspects *refer back to*, or *retrocipate on* the earlier ones. Thus each modal aspect, except for the numerical (first) aspect, contains *retrocipations*. Indeed, the meaning of any modal aspect cannot be fully grasped without an insight into its retrocipations. Conversely, we refer to the counterparts of retrocipations as *anticipations*. Not only does each modal aspect (except the first) retrocipate on the earlier aspects, but each earlier aspect (except the last) *anticipates* the later ones. In our discussion we shall be concerned only with the retrocipations and anticipations between the first four modal aspects. We shall observe that the anticipation of aspect A on aspect B is closely related to the retrocipation of B on A. These anticipations and retrocipations are also referred to as *analogies*. In keeping with our distinction between the law side and the subject side of reality, we shall find these analogies on both the law and subject side of creation. The analogies are, perhaps, even more important for scientific investigations than the "bare" modal aspects themselves. Thus, the view that the modal aspects form a sort of "layer" structure in reality, with each layer built upon the earlier ones, is prohibited. Rather than being well separated departments of reality, the modal aspects are intertwined, mutually irreducible, indispensable aspects of reality.[37] The designation and distinction of modal aspects and the exploration of their retrocipations and anticipations may be called the *fifth aim of science*.

The distinction of the modal aspects is relevant, not only to modal laws and modal subject-subject relations, but also to *typical* relationships. A typical structural law may be viewed as a typical conglomerate of relevant modal and (general) typical laws. Such a typical structural law has two limiting modal aspects, which we designate as the *founding* aspect, and the *leading*, or *qualifying* aspect. For

37. On the analogies, see Dooyeweerd C, Ch. 2.

17

example, atoms, stones, and stars are *qualified* by the physical modal aspect (we call them "physical things"), whereas plants are qualified by the biotic aspect. On the other hand, the structure of an atom is *founded* in the spatial modal aspect, since it consists of a nucleus surrounded by an appropriate number of electrons. In contrast, particles are founded in the numerical modal aspect, since they are characterized only by typical magnitudes. This intricate state of affairs, which we shall discuss in greater detail in Chapter 10, is further complicated by the fact that, within an atom, the nucleus, though itself spatially founded, functions as a particle. Such a relationship is referred to as *"enkapsis"*: the structure of the nucleus is "enkaptically bound" within the structure of the atom. In the same way, atoms are "enkaptically bound" within the structure of a molecule, and molecules within the structure of a living cell.[38]

Hence we state the modal aspects to be mutually irreducible but connected modes of *experience*, modes of *explanation*, modes of *order*, and modes of *temporal relations*. It should not be surprising to find that modes of experience and explanation are identified with modes of order and relation. In a broad sense, explanation means to order pieces of experience by relating them to other pieces of experience under a law. The modal aspects should not be understood in a Kantian sense as self-evident a priori modes of thought laid bare by a metaphysics independent of empirical science. On the contrary, our arguments for the designation of the modal aspects will be found in science (understood as the investigation of the creation), not in metaphysical speculation, based on a supposed autonomy of human thought.

1.6 *Subjects and objects*

We have now covered enough ground to justify our use of the word "subjects" to designate things which, perhaps, are more commonly referred to as "objects".[39] In fact, we prefer the linguistic use of these words, which is more original than the modern scientific and philosophical use of them.

We turn our attention to the following question: Is it possible to speak of modal, universal, biotic laws which are valid for all kinds of subjects, regardless of their typical structure? Initially, it would seem that a stone, for example, is not subject to biotic laws. In order to answer this question adequately, however, it is necessary to dis-

38. Dooyeweerd D 627 ff.
39. Dooyeweerd C, Ch. 5.

tinguish between "subjects" and "objects". Subjects are those things which are actively or directly subjected to a certain law, whereas objects, in contrast, are related to the law only passively or mediately. This implies that objects receive their creational meaning from the subject to which they are related by a subject-object relation. Thus a stone cannot be a biotic *subject*. Only organisms can be subjects to biotic laws. But atoms and molecules, rocks and sticks, may function as biotic *objects* within the sphere of some biotic law. For example, a bird's nest, as a subject, is subjected to only mathematical and physical laws (it has mathematical and physical "subject functions"). As a bird's nest, however, it can be understood adequately only as a biotic "object"; the nest has an objective biotic qualifying aspect. The bird's nest receives its true objective biotic meaning through its relation to a bird, which is a biotic subject.[40]

The distinction of subject and object is not limited to typical structures of reality. Subjects and objects also appear on the modal side of reality. A spatial point, having no extension, functions as a spatial modal object. Similarly, the path of a moving subject is a kinematic modal object since the path itself is motionless; and the state of a physical subject is a physical modal object since states do not interact. An object in some modal aspect is always a subject in another aspect. Therefore, the subject-object relation within some modal aspect is connected with the retrocipations *of* that aspect, and with the anticipations *on* it.

It is also possible to speak of subjects and objects in an epistemological context. In this case, however, only man can be a subject, since things, events, plants, and animals always remain objects of scientific or common thought. The latter can only function as subjects in an ontological context. As we have already noted, epistemology has taken priority over ontology in the dominant western philosophies. Since the Renaissance the ground motive of western thought has been the relation of freedom and nature – i.e., the relation of human thought and activity, and its natural object.[41] In developments of the past four or five centuries, the natural subjects have become increasingly objectified. Whereas they retained an independent existence, determined by their spatial extension or mechanical interaction in the philosophy of Descartes, Newton and Leibniz, natural subjects were denatured, in principle, to unknown "Dinge an sich"

40. Dooyeweerd B 42-43.
41. Dooyeweerd B, E; for the subject-object relation in humanist philosophies, cf. Dooyeweerd C 367 ff.

in Kant's thought. In modern positivistic and phenomenalistic thought they became mere appearances. Occasionally existentialistic circles have tried to restore nature in a purely individual relation of man and his environment. Parallelling this development, natural laws were reduced to mere epistemic ordering principles, whether a priori and unavoidable (Kant), merely economic (Mach), or conventional (Poincaré).

These developments are reflected in our modern terminology. Today we generally speak of natural objects, even when we discuss their subjectivity to natural law. The modern view is strongly oriented towards a completely functionalistic view of reality – i.e., the modal aspects considered as universal modes of thought are the dominant principles of explanation. In this respect, post-Renaissance philosophy differs sharply from Greek and medieval philosophies, which were usually dominated by a typicalistic view, most clearly exemplified in Aristotle's form-matter scheme.[42]

For Christian philosophy there is no need to absolutize any modal aspect, or any typical structure or relationship. At its foundation lies the acknowledgement that the creation is not independent of its Creator. On the one hand, there is no "substance" which exists independently of law, and, on the other hand, all natural subjects exist as creatures (being and becoming) under law. Because they are all subjected to laws, all subjects point to the Lawgiver: "Meaning is the mode of being of all that is created."[43] This implies that natural subjects acquire their *full* meaning only if, in addition to their subject functions, all of their object functions are also opened up in their relation to man. In this relation natural subjects receive their full religious meaning since, in his relation to God, man is the religious centre of the creation.

The distinction of subject and object enables us to achieve a clear insight into the terms "objectification" and "objectivity". In humanistic thought everything which relates to sub-human subjects is referred to as "objective". As a result, the demand for an "objective science" has acquired an entirely confused meaning. It is sometimes understood as being "intersubjective" or "public". In this case one distinguishes between individual (subjective) experience and public (objective) experience.[44] In other contexts objectivity is con-

42. Dooyeweerd C 12; Jaki, Ch. 1.
43. Dooyeweerd B 4.
44. Cf. e.g. Popper A 44ff, but also Kant A 820, B 848; for Popper, objectivity of scientific statements lies in the fact that they can be intersubjectively tested, which implies that the described phenomena should be reproducible. See also

fused with universal validity or law conformity. In the Philosophy of the Cosmonomic Idea, the meaning of the word objective is clear: *objectivity means a representation of modal and typical states of affairs referring back to earlier modal aspects.* Objectification is made possible by the existence of retrocipations on these earlier aspects, and the opening up of the latter's anticipations. The problem of objectification, which may be termed the *sixth aim of science*, shall occupy much of our attention. Spatial points, which refer back to the numerical modal aspect, enable us to find an objective numerical representation of spatial magnitudes and relative positions (cf. Chapter 2). The path of motion, referring back to the spatial modal aspect, provides us with an objective representation of the motion of a kinematic subject (cf. Chapter 4). Similarly, the state of a physical system allows us to objectify the system's interaction with other systems (cf. Chapter 5).

For physics, objectification means a representation of physical states of affairs in mathematical terms. It is frequently said that mathematics is the language of physics,[45] as if it were a merely linguistic matter. This view is erroneous and confusing. The real state of affairs is more complicated. The modal aspects, which precede the physical aspect and form the subject matter of mathematics, are universal aspects of the full creation, including physically qualified things and events. It is impossible to account for physical functioning without including the earlier aspects in one's analysis.

1.7 *The opening-process*
The three basic distinctions discussed so far define three mutually orthogonal two-fold directions. This means that our theory is intended not only to give a systematic description of contemporary knowledge of reality, but also, as an historiography, to account for the dynamic development of science. In the figure below (compare page 7) the vertex may in fact be placed in each modal aspect, except that there is no retrocipation in the first, and no anticipation in the last modal aspect.

Margenau and Park, who enumerate the following "meanings of objectivity": ontological existence ("the objective reality behind perceptible things"); intersubjectivity; invariance of aspect ("objectivity must be assigned to those properties which are, or can be made, invariant"); scientific verifiability ("Constructs which satisfy the metaphysical requirements as well as the stringent rules of empirical confirmation are called verifacts, and verifacts are the carriers of objectivity in the domain of theory"). The "metaphysical requirements", e.g., Occam's razor, economy of thought, logical fertility, simplicity, are discussed in Margenau A, Ch. 5.

45. cf. Galileo, in: Drake 237, 238.

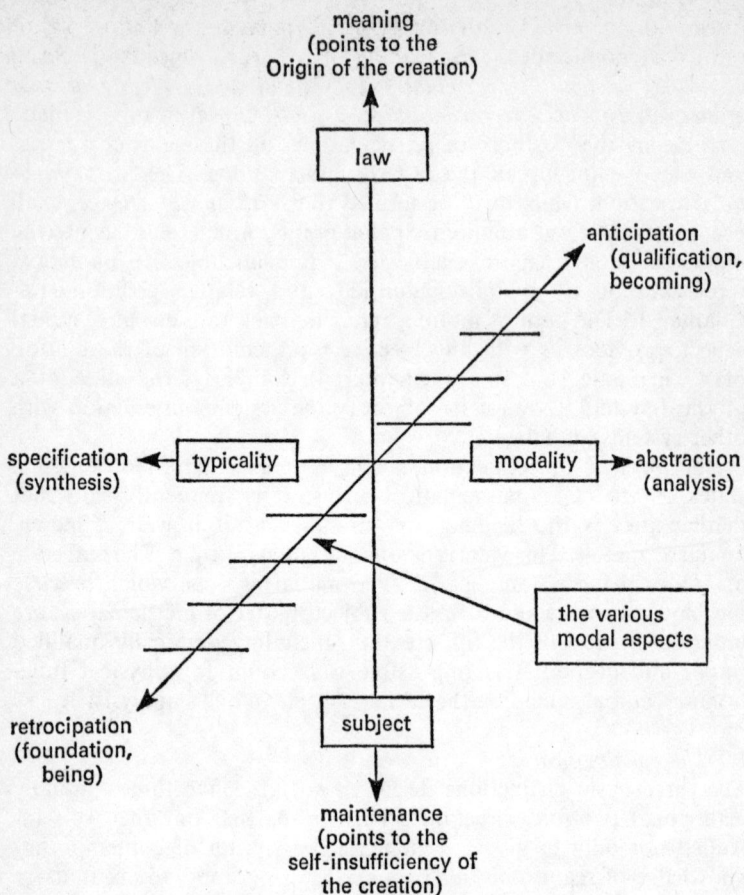

meaning
(points to the
Origin of the creation)

law

anticipation
(qualification,
becoming)

specification ← typicality
(synthesis)

modality → abstraction
(analysis)

the various
modal aspects

retrocipation
(foundation,
being)

subject

maintenance
(points to the
self-insufficiency of
the creation)

When any one of the modal aspects is recognized as a basic principle of explanation, the development of its anticipations and retrocipations is sought by means of abstraction and specification, or analysis and synthesis. This procedure is referred to as the "*opening-process*",[46] and it occurs in the "horizontal plane" (see figure), perpendicular to the vertical co-ordinate which has primarily religious meaning. We shall consider the opening-process in some detail.

46. We use this idea in a somewhat wider sense than Dooyeweerd does; cp. Dooyeweerd B 29, C 181ff, and Stafleu F, G, H, in which our views on the opening-process are applied to the 18th- and 19th-century history of electricity and magnetism.

Quite recently, several scholars in the history of science have pointed toward this opening-process. Specifically, they reject the view that ". . . scientists are men who, successfully or not, have striven to contribute one or another element to that particular constellation (of facts, theories, and methods collected in current texts) . . .", such that ". . . scientific development becomes the piecemeal process by which these items have been added, singly and in combination, to the ever growing stockpile that constitutes scientific technique and knowledge".[47]

Kuhn, in his famous book, *The Structure of Scientific Revolutions*, introduced the distinction between "normal science", which is guided by some time-honoured "paradigm", and "scientific revolutions", during which one paradigm is replaced by a new one.[48] Prior to the introduction and acceptance of any paradigm, ". . . the early developmental stages of most sciences have been characterized by continual competition between a number of distinct views of nature . . . What differentiated these various schools was . . . their incommensurable ways of seeing the world and of practicing science, in it . . ."[49] After a communis opinio is established ". . . on the assumption that the scientific community knows what the world is like . . ."[50] normal science proceeds as ". . . a strenuous and devoted attempt to force nature into the conceptual boxes supplied by professional education".[51] Eventually, in the course of normal science, anomalies, which cannot be understood within the existing framework, appear and ". . . then begin the extraordinary investigations that lead the profession at last to a new set of commitments, a new basis for the practice of science."[52]

Holton's *Thematic Origins of Science* also points to the difficulty with which new ideas are accepted. Referring to Einstein's Principle of Relativity, he observes: ". . . it is precisely such non-verifiable and non-falsifiable (and not even quite arbitrary) thematic hypotheses which are most difficult to advance or to accept. It is they which are at the heart of major changes or disputes, and whose growth, reign and decay are much neglected indicators of the most significant developments in the history of science."[53] Holton's themes are

---

47. Kuhn B 1, 2; cp. Agassi B; for an extensive discussion of Kuhn's views, see Lakatos, Musgrave; Finocchiaro.
48. Kuhn B 10, 23.
49. Kuhn B 4.
50. Kuhn B 5.
51. Kuhn B 5.
52. Kuhn B 6.
53. Holton A 190; see also Holton B, Ch. 1.

somewhat more general and less specific, and therefore more persistent, than Kuhn's paradigms. I wonder whether Kuhn would call "paradigmatic" the following themes mentioned by Holton: conservation (of mass, energy, etc.), mechanicism, ". . . macrocosmos-microcosmos correspondence, inherent principles, teleological drives, action at a distance, space filling media, organismic interpretations, hidden mechanisms, or absolutes of time, space, and simultaneity", ". . . the efficacy of geometry, the conscious and unconscious preoccupation with symmetries."[54]

For Kuhn, "A paradigm . . . is in the first place, a fundamental scientific achievement and one which includes both a theory and some exemplary applications to the results of experiment and observation. More important, it is an open-ended achievement, one which leaves all sorts of research still to be done. And, finally, it is an accepted achievement in the sense that it is received by a group whose members no longer try to rival it or to create new alternatives to it. Instead, they attempt to exploit and extend it in a variety of ways . . ."[55] Holton's themes are more or less orthogonal to the "contingent plane" of "propositions concerning empirical matters of fact (which ultimately boil down to meter readings) and propositions concerning logic and mathematics (which ultimately boil down to tautologies)."[56] "A *thematic position* or *methodological thema* is a guiding theme in the pursuit of scientific work, such as the preference for seeking to express the laws of physics whenever possible in terms of constancies, or extrema (maxima or minima), or impotency ('It is impossible that . . .')"[57] Holton also distinguishes thematic components of concepts such as force or inertia, and thematic propositions or thematic hypotheses, containing one or more thematic concepts, and which may be a product of a methodological theme.[58] As a result, Holton's themes have a more persistent character than Kuhn's paradigms: "Only occasionally (as in the case of Niels Bohr) does it seem necessary to introduce a qualitatively new theme into science".[59]

Feyerabend goes even further. Whereas both Kuhn and Holton accept the historical fact of the existence of paradigms, themes, and

54. Holton A 24, 25, 27.
55. Kuhn C 363; it is by no means easy to comprehend the meaning of Kuhn's paradigms. Masterman says that Kuhn (B) uses "paradigm" in not less than twenty-one senses.
56. Holton A 21.
57. Holton A 28.
58. Holton A 28.
59. Holton A 29; also Holton A 61ff.

normal science, Feyerabend insists that the latter is dogmatic, since it clings to a single paradigm.[60] He pleads for openmindedness, for competing views. It appears, at least from a Kuhnian perspective, that he wishes to return to the pre-paradigm period of science.[61] Feyerabend strongly attacks the "restrictive conditions" of consistency and meaning invariance, present in contemporary positivist empiricism: "Only such theories are then admissible in a given domain which either *contain* the theories already used in this domain, or which are at least *consistent* with them inside the domain; and meanings will have to be invariant with respect to scientific progress; that is, all future theories will have to be framed in such a manner that their use in explanations does not affect what is said by the theories, or factual reports to be explained."[62]

Insofar as it is assumed that sense data are independent of theories, and that the accumulation of new data cannot give rise to a change in meaning of older theories, meaning invariance is a leading motive in positivism. Criticism of this view by Kuhn, Holton, and Feyerabend is based on historical grounds. These writers give many examples which show that any change of "paradigm" implies a change in meaning, also with respect to "observational facts".

. We have argued that "meaning" is determined by the relation of law and subject, i.e., everything created has dependent meaning, as a result of being subjected to law by its Creator. Dooyeweerd formulates this in the phrase "Meaning is the mode of being of all that is created".[63] However, this does not imply another kind of "meaning invariance". Indeed, it is precisely in the opening-process that meaning is both deepened and relativized. From our perspective, we would

60. Feyerabend D 172: "Normal science, extended over a considerable time, now assumes the character of stagnation, a lack of new ideas; it seems to become a starting point for dogmatism and metaphysics. Crises, on the other hand, are now not accidental disturbances of a desirable peace; they are periods where science is at its best, exhibiting as they do the methods of progressing through the consideration of alternatives". See also Popper D and Watkins. Contrary to this, Kuhn C 364 states: "Advance from paradigm to paradigm rather than through the continuing competition between recognized classics may be a functional as well as a factual characteristic of mature scientific development".

61. Feyerabend A 320, 321: "You can be a good empiricist only if you are prepared to work with many alternative theories rather than with a single point of view and 'experience'. This plurality of theories must not be regarded as a preliminary state of knowledge which will at some time in the future be replaced by the One True Theory". See also Feyerabend E.

62. Feyerabend D 164, A 323; the latter text reads "phrased" instead of "framed". See also Bohr A 209, 210.

63. Dooyeweerd B 4.

paraphrase Kuhn's theory as follows: In the pre-paradigm phase, scientists are not yet aware of the meaning of their concepts. With the formation of the first paradigms, it is mainly the retrocipatory analogies of the modal aspects or typical structures that are discovered (this includes the search for objectivity, described in Sec. 1.6). Paradigm change is brought about by the discovery of either a new retrocipatory analogy or, even more spectacularly, by the discovery of an anticipatory analogy. Such discoveries are made possible by an increasing degree of abstraction and, simultaneously, the opening up of new typical structures, both theoretically and technically.[64]

Such developments account for the appearance of "scientific revolutions" as well as Holton's more persistent "themes". With the opening up of a modal aspect, the latter remains in existence, as a fundamental and irreducible mode of explanation, though it may be viewed in a different light. Thus, whether we use Euclidean or non-Euclidean geometries, the aim of geometry remains to account for spatial relations. The description of typical individuality structures in modern quantum physical terms requires the modal aspects with both their retrocipations and their anticipations, which, to a large extent, are describable in classical physical terms. We shall discuss several examples of this model.

Our first example is taken from the history of number theory. After the initial establishment of the meaning of number, the negative and rational numbers are introduced by abstraction. Only then are the real numbers and vectors found by anticipation. Similarly, in the history of geometry, the opened-up investigation of spatial magnitudes in the Pythagorean school was of revolutionary significance. The introduction of non-Euclidean and non-metrical space, multidimensional space, formalization of geometry are further examples of the opening up of the spatial modal aspect. In kinematics the development of both the Galilean and Einsteinian principles can be fully accounted for in our theory of the opening-process. As we shall see, the history of astronomy, of physics, and of chemistry are full of examples of sudden increases in understanding due to developments in retrocipation, anticipation, abstraction, and specification. From our vantage point, the fact that Kuhn discovers paradigmatic,

---

64. Cp. Lakatos' "research programme", Lakatos A. Lakatos' theory of research programmes, each with a "hard core" and "positive and negative heuristics" is directed to the rational understanding of the driving motive of individual scientists or groups of scientists. Hence it is not very useful for our purpose, which is the understanding of the historical development in a more general sense. Besides, Lakatos' theory lacks any insight into the creational order, to which the historical development of science is subjected.

revolutionary developments, and that Holton stresses the persistence of themes in the history of science, is understandable.

We now turn our attention to a new aspect of the problem of meaning. Does meaning change if it is opened up, and, if so, to what extent does it change and to what extent does it remain invariant? We shall show that not only does the opening-process add (anticipatory analogies) to a modal aspect, but it simultaneously influences the nuclear meaning of the aspect, together with its retrocipatory analogies. This process we refer to as *deepening and relativizing* the original meaning of a modal aspect, since in this way the aspect becomes related to later modal aspects. Our position is more complicated than either "meaning invariance" or "meaning relativism". It involves both the law side and the subject side of reality.

As an example, let us consider the concept of mass.[65] This concept was introduced first by Kepler and Galileo, but became paradigmatic only with Newton. One of the properties of mass is its conservation in chemical reactions (which, incidentally, was justified empirically long after Newton's time). This characteristic of mass is challenged in Einstein's theory of relativity (cf. Sec. 3.8). Now, one may ask whether the meaning of mass has undergone change or not? Positivists will reply that, since the factual content of the sense data related to mass has not changed, its meaning must remain invariant. Extreme operationalists will say that, as there are different experimental methods to determine mass, there are different meanings of mass which are independent of theory, and the meaning of mass will remain invariant with respect to change of theory. Feyerabend, among others, replies that any experimental method is "theory laden" and, hence, "operational meanings" are variable with theories. He states that, since mass is subject to different laws in Newtonian physics than in relativity physics, its meaning has also radically changed.[66] Still others point out that relativistic mass shares at least some of the

---

65. Cp. Feyerabend A 325ff; Kuhn B 98ff; Hesse A 64ff.
66. Feyerabend D 169: "That the relativistic concept and the classical concept of mass are very different indeed becomes clear if we also consider that the former is a *relation*, involving relative velocities between an object and a coordinate system, whereas the latter is a *property* of the object itself and independent of its behaviour in coordinate systems . . . The attempt to identify the classical mass with the relativistic rest mass is of no avail either, for although both may have the same *numerical value*, they cannot be represented by the same concept". For a similar viewpoint, see Kuhn B 101, 102.

properties of classical mass, such that some sort of family resemblance exists between the two.[67]

A view, commonly held in physics, is that Newtonian mechanics is a limiting case of relativity physics, since, at low velocities, the relativistic and Newtonian formulas become approximately equal. The relevance of this statement becomes clear only if we remember that experimental measurements always have a finite accuracy. Within given limits of accuracy, it is rather easy to determine the velocity below which it is impossible to distinguish Newtonian from relativistic results. A positivistic interpretation will say that, since in this case there is no difference between the two theories, the meaning of terms such as mass must also be the same. A realistic interpretation will insist that the meaning of mass is different in the two theories. We reject both these views.

As we shall argue in Chapters 4 and 5, Newtonian physics has a mainly retrocipatory character, whereas relativity physics concerns the kinematic opening up of the numerical and spatial modal aspects. This opening up also has a bearing on the numerical and spatial analogies of the physical modal aspect, e.g. on mass. The meaning of mass in Newtonian physics can be understood as a numerical retrocipation in the physical modal aspect. In relativity physics, this retrocipation is also opened up, inasmuch as all numerical and spatial relations become frame-dependent. But, as we have already emphasized, this state of affairs implies neither a meaning invariance (since the meaning does change), nor a loss of meaning (since it remains a retrocipatory analogy in the physical aspect). Rather, the opening-process in relativity physics results in a *deepening* and *relativizing* of the original closed meaning of mass in Newtonian physics. Relativizing does not result in a loss of meaning, especially since the retrocipatory viewpoint remains valid and useful. Indeed, there are so many instances where Newtonian mass is still relevant that it is illegitimate to characterize the Newtonian interpretation as approximately true, but formally false.

Our twofold use of the retrocipatory *and* the anticipatory, the modal *and* the typical concepts can be compared with Holton's ". . . juxtaposition of the thema-antithema . . . couple of atomism and the continuum . . ."[68] This view promises to be helpful for understanding many related problems in the history of physics, including quantum

---

67. Cf. Kuhn B 45; Hesse A 46-48, 64-65; Hesse observes that the classical and relativistic theories could not even be compared if key concepts like mass had completely different meanings in the two theories.
68. Holton A 13, 25.

mechanics. In a similar manner, we shall try to understand the problematic concepts of complementarity, wave-particle duality, correspondence (the relation of classical, modal physics to quantum, typical physics), potential and actual states, thing-like and event-like, thermodynamics and statistical physics, classical and quantum probability theory, motion and current, force and field.

It should be clear that our theory of the opening-process does not lead to "meaning relativism", which is unmistakenly present both in Feyerabend's publications and in operationalism. Both in closed and in opened-up form, meaning is bound to law. We, who study law and its relation to the subject side of reality, are similarly bound to law. We find, however, that in the opening-process not only the subject side but also the law side is involved: that is why *meaning* is opened up, and why the meaning of an opened-up modal aspect or typical structure cannot be the same as the meaning of one that is still closed.

1.8 *Science and religion*

Explicitly, we have presented the following *aims of science:* (1) the explication of laws, and (2) the reduction and deduction of laws (Sec. 1.3); (3) abstraction or analysis, and (4) reconstruction or synthesis of typical laws (Sec. 1.4); (5) the designation of modal aspects and the exploration of retrocipations and anticipations (Sec. 1.5); (6) objectification (Sec. 1.6). We could add a seventh: (7) the explanation of individual facts and phenomena.[69] We can generalize the goals of science by stating that *the aim of science is the theoretic opening up of the full creation*, as discussed in Sec. 1.7.

In addition to the theoretical opening-process, we also find many other opening-processes within the creation. There is a natural opening-process (the temporal evolutionary development of the cosmos); individual ones (the growth, flowering and decay of a plant, or the opening up of the experiential horizon of an animal or man); a technical opening-process (the opening up of possibilities laid down in the creation); an artistic one, a social one, a linguistic one, etc. In each of these cases, we expect that the four basic directions of retrocipation and anticipation, abstraction and specification will be retraceable.

The distinction of law and subject is itself directed. Subjects do not exist without laws, and via the laws they acquire *meaning* as creatures. The direction of subject-to-law points to the origin of creation, the sovereign Creator and Lawgiver, who Himself is subject to no law.

69. Popper E.

29

As viewed from the subject side, the law is the boundary of created reality, across which no subject can step. For God, the law is not a boundary,[70] but, by maintaining His laws, according to His covenant, He remains faithful to His creation.[71] Thus the direction of law-to-subject expresses the dependence of the creation upon its Creator (see page 22). The unfolding-process becomes meaningful only because of this two-fold law-subject relation.

The latter statements are clearly not of a scientific character. We have arrived at an interesting and illuminating state of affairs, which displays both the similarity and the distinction of science and religion.[72] In both cases man, who is himself a subject, searches for truth, truth about reality and about himself. In both cases the attitude of man is directed toward the origin of creation, and, therefore, in both cases, his attention is directed toward the law side of reality. The distinction of the two cases lies in the fact that in his scientific attitude, man sees the subject side "reflected" in the law side. As soon as a scientist formulates a law (finds a law conformity), he must verify it (or falsify it) on the subject side. One may even go as far as Popper who says that no law statement should be called scientific unless it is potentially falsifiable.[73]

Man, however, experiences that his scientific attitude is not sufficient for finding the full truth about reality. Through science the origin of creation cannot be found: the law side as the boundary of reality cannot be penetrated. It is in his religious attitude that man seeks to look beyond the laws. In this effort no principle of verification can help him, because any subject points to the law side and beyond for its full religious meaning. At this point human self-insufficiency becomes abundantly clear. Faithful knowledge about the origin of full reality requires revelation, the truth of which man can only find in faith. However, as we have pointed out earlier, the scientific attitude also rests on faith. The fundamental hypothesis of

70. Dooyeweerd B 99.
71. Dooyeweerd B 93.
72. By religion is understood ". . . the innate impulse of human selfhood to direct itself toward the *true* or toward a *pretended* absolute Origin of all temporal diversity of meaning, which it finds focused concentrically in itself. This description is indubitably a theoretical and philosophical one, because in *philosophical reflection* an account is required of the meaning of the word 'religion' in our argument". (Dooyeweerd B 57).
73. Popper A 41; see also Lakatos A. A similar view was already expressed by Claude Bernard, cf. Kolakowski 93. The distinction of falsification and verification reflects the law-subject relation. Scientific law statements (or "all-statements") should be falsifiable, whereas subjective existential statements (of the form "there is a . . .") should be verifiable in order to qualify as empirically meaningful. See Popper A 70.

all sciences – the hypothesis that reality is lawful – cannot be proved; it must be believed. If you don't believe it, you cannot be a scientist.

If the sovereignty of God as Creator and Lawgiver is not recognized, the unity and origin of reality must be found somewhere within temporal reality itself. In western culture, it is always man himself who is assigned the task of locating this origin, and, not recognizing the true origin, he must seek his point of reference in either one or another of the modal aspects, or in one of the typical structures. Such selection of reference points has resulted in the formation of the various mutually irreconcilable schools of philosophy, each pretending to be able to explain everything according to a single principle (see page 6). Alternatively, people may place their trust in power (economic or political), in the church, or in one of the arts.[74] Regardless of where the reference point is chosen, such a choice always leads to a dogmatic (and non-provable) over-rating of the aspect or structure concerned. A balanced and dynamic view of reality can only be achieved if the dependent and self-insufficient character of creation, of which no aspect or typical structure is overestimated or neglected, is accepted.

74. Or in astrology, superstition, myths, etc. That these convictions cannot be ruled out by their supposed lack of empirical support has been shown by Feyerabend D; cf. Kuhn B 2, D.

2. Number and Space

2.1 *Set theory and the first two modal aspects*

This book is concerned with an analysis of the foundations of physics. Such an analysis would be quite impossible, however, without taking into account the numerical and the spatial modal aspects of the creation. Therefore, we shall discuss these aspects in this chapter, though not as extensively as our discussion of the kinematic and the physical aspects in subsequent chapters. This chapter should not be taken out of the context of this book. My only intention is to investigate the numerical and the spatial modal aspects insofar as they are relevant to physics. The mutual irreducibility of these aspects will be discussed later on. In the present section I shall give a provisional outline of their meaning, and discuss their relation to set theory. The reader should keep in mind the mutual orthogonality of our distinction of law and subject, and that of the various modal aspects.

The numerical modal aspect of discrete quantity, as a universal mode of being, presupposes that every created thing is a unity, and that there exists a multitude of such unities. The numerical modal aspect is universal since there is nothing in the creation which is not subjected to numerical order. This order can be described as the order of before and after, both in its original numerical meaning of more and less, and in its analogical meaning of smaller and larger in magnitude.

The spatial modal aspect of continuous extension explains why a unique ordering of everything created is impossible solely with the numerical order of before and after. Thus different sets may have the same number of elements, and different things may have the same size. The spatial order of simultaneous coexistence (on the law side of the spatial modal aspect) makes possible the original spatial relation of relative position (on the subject side of the spatial modal aspect). This spatial modal order also involves the analogical concept of equivalence with respect to some property, thereby allowing things to share this property in different degrees. The spatial modal order is only universal if it is considered together with the numerical order. Although the order of simultaneity does not apply to everything created, we can account for all static relations if this order and the order of before and after are taken together.

Set theory is nowadays generally considered to be the basis of the theory of number. Later, in Chapter 8, I shall discuss the concept of probability and argue that it refers to the law-subject relation for individuality structures. Since the theories of probability and sets are closely related, I view the idea of a set as giving expression to the law-subject relation. Sets are always determined by some law. This is even the case with examples like "the set of all books in my room", for this refers to the law defining "books". In this context the set of all things on my desk is ill defined without further specification of a "thing". In general, classes are not identifiable, or even imaginable, unless they are defined by a law, and these laws are usually not of a mathematical kind. It is not strictly correct to say that a set is determined by a law. I prefer to say that a set has a law side and a subject side. We shall see (Sec. 2.2) that we cannot reduce the idea of a set, either to the law side, or to the subject side.

The concept of *number* cannot be studied wihout the idea of sets. Both Spinoza and Frege observed that we cannot ascribe a number to things, unless they are grasped under a genus.[1] If in the realm of concrete things and events we ignore the post-numerical modal aspects, we still have the possibility of taking some of them together in a collection. After this process of abstraction, all that remains to be said is that concrete things belong to classes of things.[2] The common property of all finite collections is that they can be counted, regardless of the spatial, kinematic, physical, etc., properties of their elements. Thus all finite collections are related either by a one-to-one correspondence, or by a one-to-one correspondence between one collection and a proper sub-set of another one. In the latter case we say that the first collection is smaller than the second one. Because this property has a universal character, we can now abstract from concrete sets. One discovers an abstract and unique collection of "natural numbers", which serve as a universal reference system for all finite collections.[3]

On the other hand we cannot talk about a set without having a previous idea of a plurality of concrete things and events,[4] nor can we dispense with the individual unity of its members. Aristotle con-

---

1. Beth A 115.
2. For the time being, I restrict myself to finite (discrete, countable) sets, to be distinguished from infinite countable sets, and continuous (uncountable) sets. We shall show that countable sets have an original, numerical meaning, whereas continuous sets have an original spatial meaning.
3. This reference system cannot be finite because of its abstract and universal character.
4. Dooyeweerd C 79ff; Kuyk; Cassirer 47-54.

sidered the individuality of things as their only property relevant to arithmetic. For Aristotle individuality meant the identity of a thing with itself and its being distinct from other things. Arithmetic had to abstract from all other properties of real things.[5] For example, the universal law of addition demands that if we add a collection of m members to a collection of n members, we always arrive at a collection of $(m + n)$ members, whatever the character of the two collections, provided they have no member in common. This implies that each member has its own subjective identity.

The concept of *space* cannot be studied without the idea of sets either. We shall see (Sec. 2.5) that the meaning of the spatial modal aspect is described by the term "extension". This means that a spatial figure is characterized by being connected and having parts. At the same time we have to consider it as an uncountable set of points, though we cannot define it as such. The fact that we can consider each spatial figure as a collection of connected and nevertheless disjoint parts is the necessary basis for the introduction of spatial magnitude.

On the other hand, the idea of a set always has a spatial aspect. In a set we have a number of coexisting members. Members can simultaneously belong to different sets. The notion of sub-sets of a set refers to the simultaneous existence of a whole and its parts. Also the concepts of "union" and "intersection" of sets refer clearly to the spatial modal aspect. In order to make the transition of all finite collections to the set of the natural numbers, one often makes use of the concept of "equivalence class". The numerical order of more and less is not directly applicable to sets, but only to equivalence classes of sets, each equivalence class uniting all sets with the same number of elements. This also shows that the spatial as well as the numerical orders are presupposed in this attempt to base a theory of number on set theory.[6] In fact, even if we talk about the *set* of natural numbers, we already refer to simultaneity.

Without the introduction of numerical and spatial orders, the subsets of a set can only be partially ordered. In order to arrive at a universal order of sets, we have to introduce the more abstract orders of numerical seriality and spatial simultaneity. For Aristotle, the number of a set was a concrete property. Frege was one of the first to recognize the abstract character of the cardinal numbers: there is only one number six, regardless of how many sixtuples of concrete things exist.[7] Even Russell's definition of the number of a class as

5. Cf. Beth A 61, 67, 68.
6. See, for instance, Beth A and Russell A.
7. Beth A 72.

"the class of all classes which are equivalent to that given class"[8] presupposes the abstraction of all properties of sixtuples, except of being classes, and having six members. It especially presupposes the abstraction from the spatial order of simultaneity, for in this case, one abstracts from the fact that so many sixtuples exist simultaneously.

It is not my intention to investigate the foundations of set theory. The above arguments only serve to make clear the mutual orthogonality of the law-subject distinction, which finds its mathematical expression in the theory of sets, and the distinction of the various modal aspects, which we intend to study in this and the subsequent chapters.

2.2 Numerical relations and the theory of groups

The numbers form an abstract reference system for any serial order. They have no concrete existence at all, and their meaning is, therefore, purely modal. I shall call them *numerical modal subjects*, being subjected to numerical modal laws (cf. Sec. 1.4). We shall investigate briefly the different number systems which are relevant to physics: the natural, integral, rational, real, and complex numbers, and vectors. We shall do this in a quasi-formal way, using a group-theoretic approach, because of the relevance of group-theory to present-day physics, and to our analysis of it.

The numerical modal aspect is not only a mode of being, experience, or thought, it is especially a universal mode of temporal relations (cf. Sec. 1.5). Time expresses itself as the numerical order of before and after on the law side of the numerical aspect.[9] The number 2 is earlier than the number 3, because the latter can be generated from the former by addition of the number 1.[10] Therefore, the number 3 "ontically presupposes" the number 2.[11]

This order can be expressed by the following law: the subjects $A, B, C, \ldots$ belong to a serially ordered collection, if there exists a binary relation R on the collection, such that

(a) either $R(A, B)$ or $R(B, A)$
(b) if $R(A, B)$, then not $R(B, A)$ – the relation is asymmetrical
(c) not $R(A, A)$ – the relation is not reflexive
(d) if $R(A, B)$ and $R(B, C)$, then $R(A, C)$ – the relation is transitive.

On the subject side, the numerical difference is correlated to this temporal order. Obviously, the statement that some number is later

8. Russell A 29.
9. Dooyeweerd A 167, 168; C 79.
10. Cp. Peano's axiomatization of the natural number system; see Russell A 15; Carnap A 38ff.
11. Popma 145.

than another one gives rise to the question: "How much later?" Indeed, the numerical difference between two numbers is related to their temporal order of earlier and later: the difference is positive or negative depending on this order (if $a > b$, then $a - b > 0$, etc.).

Since the addition of two numbers yields a number, and the difference between any two numbers is a number, the numbers form a group.[12] A *group* is a collection of distinct elements $A, B, C, \ldots$ on which an operation is defined, such that from any pair of elements A, B an element AB can be generated, according to the following rules:

(a) If A and B are elements of the group, then AB is also an element (in general, $AB \neq BA$. If $AB = BA$, the group is called commutative)

(b) $(AB)C = A(BC) = ABC$ – the group operation is associative

(c) the group contains one element E, called the identity element, such that for each element A of the group, $AE = EA = A$

(d) to each element A corresponds an inverse element A', such that $AA' = A'A = E$.

Here, the equality sign ($=$) must be understood as "is the same as", "is equal to", "cannot be distinguished from", or "can always be substituted for". There is no intrinsic way to distinguish the element AA' from the element E, for instance. The extrinsic lingual distinction only accounts for the different possibilities of generating the same element.

These four rules state the law for a group. They do not fully determine a group, however. As to the law side, one has to specify the group operation, and as to the subject side, one has to indicate the members, by stipulating a set of generators. The other members are generated by application of the group operation. Several different groups (i.e., having different members, and eventually a different group operation) may have the same group structure. In that case the groups are called different *isomorphic* models or representations of the same group structure. An isomorphism consists on the subject side of a one-to-one correspondence between the members of the two groups, and on the law side of a parallelism between the respective group operations. If the members A, B, C in one group correspond with the members K, L, M in the other group, and if $AB = C$, then $KL = M$. Thus the law does not define its subjects: the subject side cannot be reduced to the law side. Isomorphism plays an important part in the process of objectification, as we shall see in Chapter 3.

Groups may be finite or infinite. The smallest groups contain

12. Cf., e.g., Suppes A 105ff, 252ff.

just one element – evidently the identity element. For example, the number 1 forms a multiplication group, and the number 0 an addition group. They are isomorphic. The numbers 1 and -1 also form a multiplication group, consisting of just two members. Finite groups are very important in the physics of typical structures, but at present we are more interested in infinite groups.

The set of natural numbers does not form a group, though if addition is taken as the group operation, the natural numbers satisfy rules (*a*) and (*b*). But there are no inverse elements, which means that within the set of natural numbers, subtraction is not always defined. However, if we include the number zero and the negative integers, we do have a group. We generate the integers as members of the smallest addition group, which includes among its members the natural numbers. The group operation is addition, the inverse of a positive integer is a negative integer, and vice versa. To show that we have to specify some members of the group, we observe that the addition group of integers is isomorphic to the addition group of even integers, of threefolds, etc. In our approach we identify the positive integers with the natural numbers. We shall return to this matter in Sec. 2.4.

We are especially interested in relations. Within the group structure, the element AB' can be considered as expressing the intrinsic relation between two elements A and B (for short, I shall say that AB' *is* the relation between A and B). We see that the relation between two integers is their numerical difference. The reverse relation is BA'. The relation of an element to itself is $AA' = E$, the identity. Because $AE = AE' = A$, the relation of an element to the identity element is identical with the element itself. Therefore, the numerical difference between two numbers, as the basic numerical subject-subject relation, is a numerical modal subject itself.[13]

Difference is not the only conceivable numerical relation. From addition we can derive the operation of multiplication of two natural numbers (as an abstraction of the repeated addition of equally numbered collections).[14] If we introduce multiplication as a group operation, we generate the positive *rational* numbers as the members of the smallest multiplication group, whose members include the natural numbers.[15] For the group of positive rational numbers the identity

13. Cp. Cassirer 55, 56.
14. Cp. Poincaré B 8.
15. For the introduction of the set of natural numbers or the group of integers, we only need to specify one member, the number 1. All other integers are generated according to the group operation of addition. For the introduction of the multiplication group of positive rational numbers, we have to rely on the set of prime

element is the number 1, the inverse of a rational number is a fraction, and the group relation is the ratio between two rational numbers. The set of all rational numbers (positive, negative, and zero) is then defined as the addition group, whose elements include the positive rational numbers.[16] It cannot be defined as a multiplication group, because the number 0 has no inverse for multiplication.

We see that for the introduction of the rational numbers we need two group operations. This leads to the idea of a *field*, another "algebra". A field is a collection of subjects in which two commutative operations are defined (e.g., addition and multiplication), each satisfying the same rules as for groups, except that the identity element for one operation has no inverse with respect to the second operation. The two operations are connected via the distributive law: $(A + B) \times C = A \times C + B \times C$. Examples are the fields of rational numbers, of real numbers, and of complex numbers. They have the usual addition and multiplication as operations, whereas dividing by zero is not defined.

The group structure does not specify an order between the elements. For the groups of numbers discussed so far, we find that they can be ordered according to the law mentioned at the beginning of this section. As observed, we say that $A > B$, if $A - B > 0$, where "larger than zero" means "being positive". We cannot equate this order with the so-called "principle of a series". We call a set *discrete* in a certain order, if in that order each element has just one successor. Every finite collection is discrete, and so are the sets of natural and integral numbers. In a series the natural numbers (acting as "ordinal numbers") serve as indices. A set is called *denumerable* if its members can be put in such a series, i.e., if there is a one-to-one correspondence between the members of this set and the natural numbers. The order in this series is *extrinsic*, while given by the indices. We shall speak of an *intrinsic* numerical order if it is determined by the numerical values of the members themselves.

Now consider the set of rational numbers, which can be arranged in a series, as is shown in any textbook on number theory.[17] In this

numbers, and hence on the full set of natural numbers (which can only be defined with the addition as a group operation), because of the theorem that the number of prime numbers is infinite.

16. If *a*, *b*, *c*, and *d* are integers, our group-theoretical approach demands that $a/1 = a$, etc. Hence, the addition of the rational numbers must be defined as $a/b + c/d = (ad + bc)/bd$, in order to arrive at the result that $a/1 + b/1 = a + b$.

17. E.g., Courant 59, 60. Although there exists a one-to-one correspondence between the integers and the rational numbers, their groups are not isomorphic: there is no parallelism between the group operations.

series, in which a member is not necessarily larger than all preceding members, the members are arranged in an extrinsic numerical order (of the indices). If we consider the rational numbers in their intrinsic numerical order of smaller and larger, they do not form a discrete series, but a dense set. This means, in any interval there is at least one rational number, and therefore, there is an infinitude of rational numbers in any interval, and there is no empty interval, however small.

With the concept of a dense set, we have reached the limit of the closed numerical modal aspect. It is the starting point for the opening up of this aspect, anticipating later modal aspects, as we shall see presently.

2.3 *The opening up of the numerical modal aspect*
The set of line segments on a straight line having a common end point is also a group. The group operation is the spatial addition of two line segments, the inverse is a line segment in the opposite direction, the identity element is a line segment of length zero, and the group relation is a line segment equal in length to that between the non-common terminal points of two line segments. In the present context, the notions of line segment, congruence, and spatial addition are irreducible concepts: they belong to the spatial modal aspect.

Now we can introduce the real numbers as elements of the group (*a*) whose elements include the rational numbers, (*b*) which has arithmetical addition as its group operation, and (*c*) which is isomorphic to the former group of line segments. In order to make the one-to-one correspondence between the elements of the two groups definite, we have to choose an arbitrary unit segment. This shows that the set of real numbers is not identical with the set of all segments with one common end point. We could not refer to the set of all points on a line, because they do not form a group. The reference (*a*) to the rational numbers is necessary to give the reals the character of numbers. Condition (*c*) is not sufficient for this purpose. A set whose members have a one-to-one correspondence with the group of line segments is called *continuous*. There is no one-to-one correspondence possible between the elements of a denumerable group and those of a continuous group. The number of elements in a continuous group is always infinite.

The introduction of the set of real numbers as an isomorphic copy of a spatial group already indicates that the meaning of the real numbers is not originally numerical. They have an analogical meaning, anticipating the spatial modal aspect. This means that *the concept of isomorphy is a mathematical expression of the philosophical idea of analogy* (cf. Sec. 1.5). As we have seen, we may consider the negative integers and the rational numbers as expressing modal numerical relations between natural numbers and among themselves, and thus as modal abstractions of relations between discrete collections. So the modal meaning of negative and rational numbers remains completely within the closed numerical modal aspect of discrete quantity.

The denseness of the set of rational numbers allows us to open up this meaning, for instance in the following way. The set of rational numbers contains Cauchy sequences: infinite sequences of elements A_n, given according to some law, such that for any positive number e (however small) there is a number N, such that if $n > N$ and $m > N$, then $| A_m - A_n | < e$. It is assumed that such a sequence has a limiting value A, meaning that, for every positive number e, there is a number N such that if $n > N$, then $| A - A_n | < e$.[18] It may be observed that the existence of this limit does not depend on an actual completed infinitude of the series as a totality: an infinite discrete set does not have a last member.

Even if a set is dense, it may occur that the limit A of a Cauchy sequence is not a member of the set. There are in fact Cauchy sequences of rational numbers whose limits are not rational numbers themselves. The set consisting of Cauchy sequences of rational numbers is the set of all real numbers. We say that the inclusion of these limits *completes* the dense set of rational numbers, making it a continuous set of real numbers.[19]

Actually, the real numbers cannot be defined in this way. For instance, it is already presupposed that the limit A of a Cauchy sequence of rational numbers is a number, because otherwise the numerical difference $| A - A_n |$ would have no meaning.[20] However, for the same reason, it is objectionable to say that this limit is not a number. It is an assumption to state that the limits of Cauchy sequences of rational numbers are (real) numbers, and one has to show that this assumption is warranted.

According to Dooyeweerd,[21] rational and real numbers must be

18. Courant 39, 40, 60.
19. Up till the end of the 19th century, the distinction between denseness and continuity was not clearly recognized, cf. Grünbaum D 13. In the past, continuity was sometimes defined as "infinite divisibility", but this leads only to denseness.
20. Boyer 284-290. To avoid this pitfall the modern approaches of Weierstrass, Cantor, Dedekind, and Russell have been institutionalized.
21. Dooyeweerd C 79, 88, 170ff, 383; see also Strauss. In fact, this is not quite a new view: for some time, the negative numbers were called "numeri absurdi", "aestimationes falsae" or "fictae", the irrational numbers "numeri surdi", and the complex numbers are still called "imaginary"; cf. Beth B 72, 73.

considered mere "functions of numbers", the only original numbers being the natural numbers. I agree that the natural numbers are primitives, whereas the existence of rational and real numbers depends on the existence of natural numbers. Nevertheless it is meaningful to speak of *numbers*, also in the case of negative, rational and real numbers, as *modal subjects to numerical laws*. In order to see this, we have to recall that the mutual relationship of law and subject plays a prominent part in our theory: we cannot have laws without subjects, or subjects without laws. But it may be imagined that mankind first discovered certain subjects (e.g., the natural numbers) and some laws (the laws of addition and multiplication) to which they are subjected. Afterwards, one discovered other laws (subtraction, division) pertaining to the same subjects. But then one also discovered other subjects (negative and rational numbers) to the same laws. In our view there is no reason to call these newly discovered subjects mere functions of the already known primitive subjects.[22] The real numbers are also subjected to the same numerical laws of addition, multiplication, subtraction and division, as the rational numbers.[23] Thus these numerical predicates of infinite sets of rational numbers behave as subjects to numerical laws.

But we still have to question the meaning of these numbers. As observed, the meaning of the rational numbers remains completely within the closed numerical modal aspect, because they denote numerical relations between discrete collections. The set of all real numbers turns out to be non-denumerable, i.e., it is impossible to find a one-to-one correspondence of this set with the set of natural numbers. The meaning of a non-denumerable set cannot be found in the closed numerical modal aspect. But this meaning is found as soon as we discover the one-to-one correspondence between the set of all real numbers and the set of line segments introduced above. Hence, the meaning of the set of real numbers anticipates the spatial modal aspect.

This is also the case with the meaning of individual real numbers. Real numbers do not refer to discrete sets, but to continuous sets, such as the set of all points on a line. Moreover, real numbers objectify magnitudes, first of all spatial magnitudes: lengths, areas, volumes. It was the great discovery of the Pythagorean School, that the rational numbers are insufficient for the numerical objectification of spatial magnitudes. The diagonal in a unit square has a length of

22. Beth A 155.
23. Cf. Beth B 50ff.

$\sqrt{2}$, and it can easily be shown that this is not a rational number. In order to represent such magnitudes, one needs the real numbers.[24]

We find, therefore, that the meaning of the real numbers anticipates the later modal aspects. This explains their potential character. The limit of an infinite series is never actualized, but in the retrocipatory direction, real numbers become actual. The length of a line segment is an actual, real magnitude. In the opening-process the original closed meaning of a modal aspect is deepened and relativized. The deepening means that not only discrete sets, but also magnitudes can be numerically ordered. With real numbers, non-numerical subjects can be ordered according to their magnitude without gaps or holes. This relativization of modal meaning entails the loss of the discrete or denumerable character of number in the numerical modal aspect.

2.4 *Vectors*

In Sec. 2.2 we introduced the negative and the rational numbers with the help of group theory, and we observed that their meaning depends on the meaning of the natural numbers: they refer to numerical relations between concrete collections. Apparently for this reason some mathematicians[25] introduce the integers and rationals as equivalence classes of differences or ratios between natural numbers. Thus the integer 2 is the equivalence class of all differences $(2 + b) - b$, where b ranges over all natural numbers. In this view the positive integers should not be identified with the natural numbers, as we did, and, depending on the context, the symbol "2" may stand for a natural number, an integer, a rational number, and eventually for a real or a complex number. This view is understandable if one considers the numbers as logically definable. In our view, numbers are discovered and are modal subjects under a law. Therefore we have no difficulty in identifying the number 2 as being the same member in different sets. I call this the *modal identity* of a number.

The time order in the closed numerical modal aspect is that of earlier and later, and we call two numbers *equal* if they have the same position in this order. Therefore we allow only one number 2, whether understood as a natural number, an integer, a rational, or a real number. However, if we apply the abstract numerical order of smaller and larger to concrete subjects or collections two or more subjects may be equivalent with respect to some property.

In that case there will be at least one other property with respect to which they will be different. In many cases it will be possible to order

24. Beth D 77ff; A 23ff.
25. Beth B 34ff; Russell A, Ch. 7.

a set of subjects according to two or more independent properties. Thus we have series with two, three, or more indices. Discrete series can always be ordered in a single numerical order, but this is not always desirable. It might also be that two independent properties have a continuous "spectrum", in which case a unequivocal single numerical order is impossible. This notion of independence anticipates the spatial order of simultaneity, and therefore we have an opening up of the numerical modal aspect on the law side.

We shall speak of magnitudes as non-numerical relations, which can be objectified by real numbers. There are non-numerical relations which can only be ordered in a serial order of smaller and larger, if they are decomposed into components, which simultaneously determine these relations. This applies in the first place to spatial position, but also to force, velocity, the physical state of a system, etc. Such relations are not objectified by a single real number, but by a multiplet (or n-tuple) of real numbers, a so-called *vector*. The minimum number of reals needed for an objectification of a property or relation is called the latter's *dimension*. The corresponding vector has an equal number of independent components, which is, therefore, sometimes called the vector's dimension.

By way of example, and because of their relevance to physics, we shall briefly review the theories of vectors, complex numbers, and Hilbert space.

We define a *vector* as an n-tuple of n real numbers, written as
$$\mathbf{a} = (a_1, a_2, a_3, \ldots, a_n)$$
subjected to the following rules:
– the sum of two vectors is a vector defined as
$$\mathbf{a} + \mathbf{b} = (a_1 + b_1, a_2 + b_2, \ldots, a_n + b_n)$$
– the product of a vector with a real number c is a vector defined as
$$c\mathbf{a} = (ca_1, ca_2, \ldots, ca_n)$$
Introducing the n unit vectors
$$(1, 0, 0, \ldots, 0), (0, 1, 0, \ldots, 0), \ldots (0, 0, \ldots, 1)$$
any vector can be written as
$$\mathbf{a} = a_1(1, 0, 0, \ldots) + a_2(0, 1, 0, \ldots) + \ldots + a_n (0, 0, \ldots, 1)$$
The vectors with the same number n of components form a group with vector addition as group operation, and the zero vector $(0, 0, 0, \ldots, 0)$ as identity element. The inverse of a vector $\mathbf{a}$ is $-\mathbf{a} = (-1)\mathbf{a}$. It is easily verified that the set of real numbers is isomorphic to the set of one-component vectors. The independence of the components is not changed by addition.

Next we define the *scalar product*, a functional of two vectors having the same number of components, as the real number
$$\mathbf{a}.\mathbf{b} = a_1b_1 + a_2b_2 + \ldots + a_nb_n$$

Because the result is a number, not a vector, this product does not define a group. The *norm* $|\mathbf{a}|$ of a vector $\mathbf{a}$ is defined by

$$|\mathbf{a}|^2 = \mathbf{a}.\mathbf{a} = a_1^2 + a_2^2 + \ldots + a_n^2$$

We take the square root of this scalar product, not the product itself, in order to find that the norm of a one-tuple is equal to the absolute value of its only component, and for other vectors, that the norms of vectors which are multiples of one another are additive.

One may wonder whether there exists an operation analogous to multiplication that gives rise to a *field* of vectors. This is indeed the case with the two-component vectors called *complex numbers*. They are often written as

$$a_1 + a_2 i = a_1(1, 0) + a_2(0, 1) = (a_1, a_2)$$

Here the vector $(1,0)$ is identified with the real number 1, and the vector $(0,1) = i$ is the so-called imaginary unit. The addition of complex numbers is defined above. We call $\mathbf{a}^* = a_1 - a_2 i$ the complex conjugate of $\mathbf{a} = a_1 + a_2 i$. The complex conjugate of a "real number" $(a_1, 0)$ is identical with itself. The product of two complex numbers is defined as the complex number

$$(a_1, a_2)(b_1, b_2) = (a_1 b_1 - a_2 b_2, a_2 b_1 + a_1 b_2)$$

Together with the addition, this defines a field. The unit vector is $(1, 0)$, and the multiplicative inverse of $\mathbf{a}$ is $\mathbf{a}^*/\mathbf{a}.\mathbf{a}^*$. We see that $i^2 = -1$, according to the popular definition of i.

Because of its relevance to physics we recall that the complex numbers can also be represented in other ways by a pair of real numbers. The most important is the representation in terms of sine and cosine functions, or equivalently, as an exponential function. If

$$a = C \cos x, \text{ and } b = C \sin x$$

then $a + bi = (a, b) = C \cos x + Ci \sin x = C \exp ix$

The norm of this complex number is $|C|$, and x is called the phase of the complex number. For any integer n,

$$C \exp i(x + n.2\pi) = C \exp ix$$

This representation is especially convenient with respect to multiplication:

$$(C \exp ix)(D \exp iy) = CD \exp i(x + y)$$

The solutions of many problems concerning functions of real numbers are only possible, or more easily obtained, if the latter are considered as vectors $(a, 0)$ – i.e., if we consider those functions as functions of complex numbers.[26] This shows that the full meaning of opened up modal subjects (real numbers) becomes clear only if the law side is also opened up (by the introduction of vectors). Besides vectors, there are other structures, like tensors and matrices, in which

---

26. Beth B 42.

each component has two or more indices. They anticipate more complicated spatial or non-spatial relations than vectors are capable of doing. With the introduction of real and complex numbers it is also possible to anticipate the kinematical and later modal aspects, e.g., as in integral and differential calculus.[27]

The concept of a vector can be further extended. We can consider vectors with complex components, and we can multiply these vectors with complex numbers. Next we can consider functions of real or complex variables instead of vectors, and multiply them either with real or complex numbers. In quantum physics one makes use of a so-called *Hilbert space*, which is not a space (there are no spatial subjects in it), but a set of complex functions, anticipating the spatial and later modal aspects. Here it is not immediately necessary to define exactly the character of the scalar product (which can be different for different cases), if only the functions belonging to the set and the scalar product conform to the following rules.

If a and b are arbitrary complex numbers, and f_1, f_2, and f_3 are arbitrary members of the set,

- $g = af_1 + bf_2$ is also a member of the set, which is therefore a group under addition;
- there exists a functional (f_1, f_2) called the scalar product, which is a finite complex number;
- $(f_1, f_2) = (f_2, f_1)^*$
- $(af_1, bf_2) = a^*b(f_1, f_2)$
- $(f_1 + f_2, f_3) = (f_1, f_3) + (f_2, f_3)$, and
  $(f_1, f_2 + f_3) = (f_1, f_2) + (f_1, f_3)$: the scalar product is a linear functional;
- the norm $||f||$ of the function f is a real non-negative number defined by $||f||^2 = (f, f)$
- if $(f_1, f_2) = 0$, we call f_1 and f_2 orthogonal, which implies that they are mutually independent. There exists a maximum number m of mutually independent and normalized functions $n_1, n_2, \ldots, n_m$, such that
  $(n_i, n_i) = 1$ for $i = 1, 2, 3, \ldots, m$
  $(n_i, n_j) = 0$, if $i \neq j$ for $i, j = 1, 2, 3, \ldots, m$
- this implies that any function f in the set can be written as
  $f = a_1 n_1 + a_2 n_2 + \ldots + a_m n_m$
  where $a_1, a_2, \ldots, a_m$ are complex numbers, $a_i = (f, n_i)$;
- with respect to the "basis" (the set $n_1, n_2, \ldots, n_m$) f can be written as the vector $f = (a_1, a_2, a_3, \ldots, a_m)$. The basis is not unique. In fact, there is an infinitude of possible bases for a Hilbert space.

27. Beth B 67.

The possibility of "mapping" a Hilbert space on a set of vectors means that all Hilbert spaces with the same value for m are isomorphic to each other.[28] This number m, the dimension of the set, may be finite (as assumed above), infinite, and even non-denumerable.

The theory of vectors as discussed in this section could be called a pure vector theory in as far as it does not contain retrocipations to the numerical modal aspect, but only numerical anticipations of this aspect to later modal aspects. Therefore, this theory belongs to opened-up number theory. It should be clear, especially from the last few paragraphs, that the opening-process goes hand in hand with further abstraction. Later we shall see that it is also extremely important for the study of structures of individuality.

2.5 The spatial modal aspect

As we have seen in Sec. 2.1, an initial abstraction from the typicality and individuality of concrete things and events requires that we take some of them together into collections and determine their number. We then observe that concrete things have several more or less static properties which they share with other things. Thus we find that several things may be similar by having a colour. Subsequently, we find that within the set of subjects having a colour there are equivalence classes of subjects having the same colour. Therefore, the common property of being coloured implies the distinguishing property of having different colours. We are especially interested in properties, which allow us to order subjects serially. But no property leads to a unequivocal serial ordering because there are always subjects which are equivalent with respect to that property. Most properties of this kind, which allow us to compare subjects, are of a typical character. Therefore, we need to look for a universal relation between subjects – a relation to which all these properties may retrocipate because it has the same structure. This leads us to the spatial modal aspect.

The second modal aspect can be characterized by the words "spatial extension".[29] If subjects are abstracted from all but the numerical and the spatial aspects, they become static, spatially extended figures. That a spatial subject is extended means that it is *connected* and that it has *parts*.[30] (Discrete sets have merely uncon-

28. Jauch 24.
29. Dooyeweerd C 85ff.
30. The relation between a whole and its parts, the terms "internal" and "external", "inclusion", "exclusion", "overlap", etc. have an originally spatial meaning; they return in later modal aspects with an analogical meaning, referring back to the spatial modal aspect. According to Dooyeweerd (C 453-456), the *logical* relation of a whole and its parts is a numerical retrocipation. But this relation is invariably subjected to the spatial order of simultaneity, and can there-

nected sub-sets). Thus being extended not only has a subject side (having parts), but also a law side (being connected).[31] In the simplest case of a linear manifold, we say that it is everywhere connected if there are no holes or gaps. In more-dimensional subjects there may be holes or gaps, but the parts of the subjects must be directly or indirectly (via other parts) connected, otherwise we do not consider it one spatial subject.

We say that there is a spatial relation between two subjects if there is at least one third subject of which the two subjects are parts. In this way we define an n-dimensional manifold as the set consisting of all n-dimensional subjects which are spatially related. This idea of a manifold can also have an analogical meaning – namely in the case of a property (such as colour) displaying a continuous spectrum. Usually one assumes that the colour of a subject is objectified by a point in this spectrum. But modern physics has shown that the colour is more adequately represented by an extended part of this spectrum, or rather by a numerical distribution over this spectrum. It seems that this is not only the case with colour, but with every kinematic or physical property displaying a continuous spectrum.

The spatial aspect precedes the kinematical, and depends on the numerical. The dimensionality of space, on the law side, and spatial magnitude (length, distance, area, volume), on the subject side, retrocipate on the numerical aspect. In both cases we find reference to the opened-up numerical aspect: the concept of dimension refers to the concept of independence of the components of a vector, and spatial magnitudes have real numerical values. This does not mean that before the discovery of real numbers people did not know what length was. But after that discovery, their understanding of length and other spatial magnitudes was deepened.

From a modal point of view, the number of dimensions of spatial subjects is not limited to three, as is the case for physically qualified subjects. Every attempt thus far to give a geometrical explanation of the latter state of affairs has failed.[32] It appears that we must accept this as one of the most general *typical* laws concerning physically qualified individuals. This law is irreducible to any (at least spatial) modal law.

Because the Greeks experienced difficulties comprehending irra-

fore only be a spatial retrocipation. The logical relations of implication, conjunction, and disjunction, and the principium contradictionis are also spatial retrocipations in the logical modal aspect.

31. Beth D 77ff.

32. Poincaré B 84; Beth D 128; Jammer A 172-184; Vuillemin 170-176; but see also Weizsäcker B 258.

tional numbers they formulated a purely geometric theory of space – Euclidean geometry. The first opening up of the spatial modal aspect was thus in the direction of abstraction: the introduction of points, lines, planes, and idealized spatial subject or objects, and the irreducible laws (or axioms) to which they are subjected. This abstraction was necessary for the subsequent study of many kinds of spatial structures, such as triangles, polygons, and the regular solid bodies. The next opening up of the spatial modal aspect occurred in the retrocipatory direction with Descartes' discovery of analytic geometry. Anticipatory opening up can be recognized in the introduction of non-Euclidean geometries, the application of group theory to spatial structures, the study of non-standard topologies, affine geometry, etc.

2.6 *The spatial subject-object relation*

The distinction of subjects and objects, made in the Philosophy of the Cosmonomic Idea, can best be illustrated with respect to spatial objects and magnitudes. (cf. Sec. 1.6). In the spatial modal aspect, objects have the character of boundaries. The proper parts of a spatial subject cannot have more or less dimensions than the subject itself. A two-dimensional subject can only have two-dimensional parts. Just as we can only add collections which have no members in common, we can only add magnitudes of spatial subjects which have no parts in common. But they may have common boundaries, because the boundaries are not parts of the subject. A boundary of a subject always has a lower dimension than the subject itself, and, therefore, its subjective extension (with respect to the magnitude of the subject) is zero (it has "measure zero"). Spatial boundaries have an objective meaning within the spatial modal aspect. They delimit the objective magnitude of the subjects, and they allow the introduction of numerical ordering within the spatial aspect.

The simplest spatial objects are points, which have no spatial extension at all. Still they have an important spatial meaning: as the boundary of a line segment, a spatial point serves to determine its length – i.e., an objective magnitude of the line segment. Similarly, in two-dimensional space, a line segment can only function objectively, as a boundary of a triangle – e.g., by determining its area, which is again an objective spatial magnitude and refers back to the numerical modal aspect. Thus we find that the spatial modal aspect is the first aspect to have objects as well as subjects.[33]

33. Dooyeweerd C 383ff; Dooyeweerd's statement that an object in some modal aspect cannot be a subject in the same modal aspect is obviously wrong.

It is of no use to *define* a line, a plane, or a space as collections of points, lines or planes.[34] To be sure, we have a continuous, non-denumerable collection of points on a line, but this cannot serve as a constitutive definition. Rather, the line constitutes the collection of points. Collections of this kind have a dependent, a so-called ana-logical, meaning. This becomes apparent if one tries to assign a num-ber to a collection of points such as a line segment. It can easily be proved that there exists a one-to-one correspondence between the points of this line segment, and to the points of any other line seg-ment, regardless of their relative length. Therefore, length, as an objective magnitude of the line segment, has no relation whatsoever to the number of points on the line segment.

Spatial points have no subjective dimensional existence. Their existence depends on the subjective existence of spatial subjects like line segments. Therefore, in Sec. 2.3, we did not relate the set of real numbers to points, but to the set of line segments with one common end point on a straight line. These line segments form a group with spatial addition as the group operation. We cannot define the col-lection of points as a group: the addition of points does not yield a point and the distance between two points is not a point, but the magnitude of a line segment. However, the group character of spatial addition and its isomorphic relation with numerical addition makes it possible to objectify the former, and in this respect the end points of the line segments have an objective meaning with respect to the group.

As we shall see, we use a coordinate system to objectify points in an n-dimensional space, attaching a set of n real numbers (a vector) to each point. In a similar way, we objectify lines, planes, etc., in three-dimensional space, e.g., with the help of a numerical function. Con-sider now an $(n - 1)$-dimensional boundary in an n-dimensional space, described by a continuous function $f(\mathbf{r}) = 0$, where $\mathbf{r}$ denotes the vector, ranging over all points in the n-dimensional space. All points on one side of the boundary are characterized by $f(\mathbf{r}) > 0$, and all points on the other side by $f(\mathbf{r}) < 0$. This shows that the concept of a boundary (a spatial object) refers back to the numerical order of smaller and larger. It enables us to order all points in n-dimensional space in a quasi-serial order, because it is a many-to-one correspondence: many points correspond to a certain numerical value $f(\mathbf{r}) = a$.

34. Cp. Suppes B 310.

2.7 *Spatial relations*

We say that there is a spatial relation between two subjects if they are bound together in a common spatial manifold. Thus the spatial order is *co-existence, static simultaneity,* or *equivalence,*[35] and the corresponding subject-subject relation is *relative spatial position.* In the kinematic modal aspect simultaneity has only a limited, analogical meaning, as is shown in the theory of relativity (cf. Ch. 4), whereas in the numerical order of before and after simultaneity is absent. With respect, therefore, to the quasi-serial order discussed in Sec. 2.6, all points with vector $\mathbf{r}$, such that $f(\mathbf{r}) = a$, are equivalent. They simultaneously lie in the same $(n - 1)$-dimensional manifold objectified by this equation.

Just as numerical relations are subjected to a serial order (cf. Sec. 2.2), spatial relations are subjected to an order of equivalence. We speak of an equivalence relation R over a set if for any two elements A and B of the set we can say that either $R(A, B)$ or not, and

(a) $R(A, A)$ for all A – the relation is reflexive

(b) if $R(A, B)$, then $R(B, A)$ – the order is symmetric

(c) if $R(A, B)$ and $R(B, C)$, then $R(A, C)$ – the order is transitive.

All elements which are equivalent with a certain element A constitute the equivalence class of A. It is a sub-set of the whole set over which the equivalence relation R is defined.

It can be shown that if this is the case there must be some property by which different equivalence classes in the same set can be distinguished. Consider the relation "parallel to" for straight lines in an Euclidean space. The equivalence classes of parallel lines can be distinguished by their relative direction. But his distinguishing property does not necessarily lead to a numerical order.

One may wonder whether the order of equivalence is not prior to a serial order, especially because mathematicians like Russell and Beth use the concept of equivalence in order to derive the concept of number (cf. Sec. 2.4). However, in many, if not all, cases the concept of equivalence presupposes numerical order, because either equivalence classes of similar kinds can be serially ordered, or within an equivalence class we find a serial order of another kind. Furthermore, as we have seen, there is no equivalence, but only equality, within the number system itself. And, finally, the concept of equivalence presupposes plurality, because we can only meaningfully speak of equivalence with respect to at least two subjects.

In which ways can spatial figures differ or be equivalent? Generally

35. Dooyeweerd A 166; C 85; Leibniz already considered space and time as orders or arrangements of co-existing and successive things or phenomena. Cf. Jammer A 4, 115; Whiteman 383; Čapek 15ff.

speaking, by their shape, their magnitude, and their relative position. If two subjects have the same shape we call them similar. If they also have the same magnitude (area or volume) we call them congruent The concept of magnitude refers back to the numerical modal aspect and, more specifically, to the operation of addition: if we take two disjoint subjects together, we have to add their magnitudes. The concept of similarity is an equivalence relation, but it clearly does not lead to a universal ordering of spatial subjects. The concept of magnitude allows us to find such an order, but this has a numerical, not a spatial character. Only spatial position can be qualified as an irreducible, universal, spatial subject-subject relation.

If two subjects are congruent, they can only differ in their position because otherwise they must be identical. Two subjects may have parts in common, they may have nothing more than a boundary in common, or they may be completely disjoint. Otherwise, it is difficult to use the concept of relative position (although it is probably intuitively clear) without an objective description – namely, the distance and relative orientation of two subjects. The shape of a subject is also determined by the relative position of its boundaries, just as is its magnitude. Relative position is subjected to the order of equivalence: the subjects considered should have the same dimension, and must be in the same manifold – these are equivalence relations.

Spatial figures can be objectified by their boundaries, in the simplest cases by spatial points – e.g., a triangle by its vertices. If the shape of a subject is given, one needs n points to objectify the position of an n-dimensional subject in an n-dimensional manifold. As a consequence, the relative position of two subjects is objectified by the distances of corresponding pairs of such points. This determines the relative distance as well as the relative orientation of the subjects. Thus the distance of two spatial points (besides the angle between two lines) is an objective, spatial relation.

In Euclidean geometry, the relative position of points is found with the help of a Cartesian coordinate system, which allows us to represent each spatial point by a vector $(x, y, z, \ldots)$. If we have two points characterized by the vectors $(x_1, y_1, z_1, \ldots)$ and $(x_2, y_2, z_2, \ldots)$, the difference vector $(x_1 - x_2, y_1 - y_2, z_1 - z_2, \ldots)$ characterizes the relative position of the two points. The distance of the points is the norm d of this vector, determined by

$$d^2 = (x_1 - x_2)^2 + (y_1 - y_2)^2 + (z_1 - z_2)^2 + \ldots$$

This expression is called the *metric* of Euclidean space. By a metric we understand a law according to which a numerical value can be assigned to a non-numerical property or relation (cf. Ch. 3). The above formula is an objective representation of this law for the determination of lengths and distances in Euclidean space.

2.8 *Objectivity in the choice of coordinate systems*

The Euclidean metric defined above is independent of the choice of the Cartesian coordinate system – i.e., it is not affected by any translation (or displacement), rotation, or inversion of the latter. We shall discuss this statement because the natural sciences claim to be objective, and because its relevance is called into question by modern conventionalist authors.

The possibility of assigning real numbers to points on a straight line depends on the one-to-one correspondence between the numerical addition group of real numbers and the spatial addition group of line segments on a straight line. This correspondence is not unique in two senses: we are free to choose a unit (see below), and we are free to choose the common end point of the set of line segments. We require that the distance between two points (the objective relation between two spatial subjects) be independent of this arbitrary choice. This is expressed by saying that the distance is invariant under translations of the coordinate system: the space is *homogeneous*. All possible displacements form a group, isomorphic to the group of all spatial difference vectors.

We are also free to choose a unit of length – that is, if we have chosen a zero point, we are still free to choose a point to which we assign the number 1. This arbitrariness is limited by the requirement that the distance between two spatial points be independent of rotations of the coordinate system around any axis and about any angle. This is called the *isotropy* of space. This implies that the unit be the same along all coordinate axes. The set of all possible rotations in a plane forms a commutative group. Rotations around different axes in more-than-two-dimensional space form a non-commutative group.

If we have chosen a set of coordinate axes and a unit, we are still free to assign the plus and minus directions on each axis. This results in inversion symmetry, the operation under which the distance must be invariant. The rotations together with the reflections form the full orthogonal group. Each finite translation or rotation can be obtained as the result of a continuous motion. However, this is not the case with inversion, which refers back to the numerical order of before and after. This implies that it will not always be possible to bring congruent spatial figures to coincidence merely by a combination of translations and rotations. For example, the right- and left-hand gloves of a pair cannot replace each other.

If we change the unit, all distances are changed by the same ratio. All possible transformations of the unit form a multiplication group which is isomorphic to the multiplication group of positive real numbers. Therefore, if we change our unit, all distance ratios must remain the same. We feel intuitively that the distance should be geometrically

52

independent of the choice of the unit of length, but this cannot be accounted for by a numerical analysis alone. In the theory of vectors, as discussed in Sec. 2.4, there is nothing of this kind: units do not occur in number theory. The meaning of the spatial subject-subject relation is determined by the irreducible meaning of the spatial modal aspect, and cannot be reduced completely to the numerical relations which objectify spatial relations. From an arithmetical point of view, the replacement of the metre by the centimetre as a unit of length causes all distances to become a hundred times larger. Transformations of this kind are sometimes called "trivial", but they are not since they express the mutual irreducibility of the numerical and the spatial modal aspects (see also Secs. 3.5 and 4.6).[36]

These invariance properties are not only relevant to distances, but also clarify the concepts of congruence and similarity. We say that two spatial figures (irrespective of their relative position) are congruent if we can transform the one into the other by an operation belonging to the full group of translations, rotations, and inversion. We say that two figures are similar (have the same shape) if besides such an operation we may also multiply all linear dimensions of one figure by a real number in order to arrive at the same result. This implies that if two figures are congruent or similar, they remain so under any transformation of the coordinate system of the types discussed here.

The standard Euclidean metric is invariant under translations, rotations, and inversion of the coordinate system. In contrast, one can show that any other metric singles out a particular point, line, plane, or direction. Thus we can say that the standard metric represents the isotropy and homogeneity of space, which are assumed here because only spatial *relations* between subjects are relevant, and not the "absolute position" of any subject.

The metric is only dependent on the choice of the unit. This arbitrariness reflects the "amorphousness" of space, by which we mean that we cannot assign a certain "amount" of points to a certain line segment. In fact, a one-to-one correspondence is possible between the points of any pair of intervals, irrespective of their relative lengths. Therefore, the length of an interval as expressed by a certain number, is not an intrinsic spatial property. This is properly stressed by Grünbaum in his extensive studies on the alleged conventionality of

36. The arbitrariness of the choice of the unit, sometimes called "gauge invariance" must not be confused with the so-called "magnitude invariance", according to which many properties of, e.g., spatial figures only depend on their shape and not on their magnitude. The former invariance is universally valid while the latter has a far more limited validity. In particular, it is false for typical relations, such as the size of atoms. Cf. Čapek 21-26.

the metric.[37] Grünbaum is the main contemporary (though moderate) proponent of conventionalism. He repeatedly refers to Poincaré and Riemann, but, in fact, conventionalism is merely a modern form of nominalism, which has its roots in the late Middle Ages and was defended by Berkeley and Mach.[38] Grünbaum uses the amorphousness of space as an argument for the equivalence of all conceivable coordinate systems, but does admit that some coordinate systems (especially the Cartesian) are more convenient than others. This argument appears to lead him astray because he conceives space exclusively as a continuous point set. Although any manifold is indeed a point set, we do not believe it can be *generated* by points because they are objects rather than subjects. Points can only be identified if they refer to one or more spatially extended figures. This identification is a prerequisite for any discussion of coordinate transformations. If we apply transformation rules to numerical vectors, we simply get other vectors. However, in the spatial case we assume that the same point, if considered with respect to different coordinate systems, can be represented by different vectors. Here we have another example of *modal identity* (see page 42). "Space" is not an actually existing concrete reality, but a mode of being of spatial subjects with their spatial subjective extension and relations.

In the non-standard metric of a semiplane discussed by Grünbaum, the distance is not invariant under a translation of the coordinate system along the y-axis.[39] The non-standard metric which he discusses elsewhere[40] is not invariant under rotations of the coordinate system. As Grünbaum rightly observes, the assignment of real numbers to spatial points only effects a coordinatization, not a metrization of the manifold.[41] However, we suggest that his non-standard metrizations do not define *proper* spatial subject-subject relations. Although a third spatial subject (the coordinate system) is used to objectify the spatial relations between two subjects, we require a metrization which keeps this spatial relation independent of the position of that third subject. This is a *requirement of objectivity* which presupposes the homogeneity and isotropy of space, that is, rejection of any absoluteness of space with respect to position or direction.[42]

37. Grünbaum D 12, 13.
38. see Kolakowski, Ch. 2 and 6. For a critique of conventionalism, see Popper A 78ff, 144ff; Friedman.
39. Grünbaum B 18ff; D 16ff.
40. Grünbaum B 98ff; Grünbaum, in: Henkin 204-222.
41. Grünbaum B 16; D 34.
42. It should be noted that my critique is not quite appropriate to Grünbaum's alternative metrization mentioned in note 39 supra. His semi-plane is only sym-

This does not mean that other metrizations should be rejected in all circumstances. Often they are very useful (e.g., polar coordinates for spherical-symmetrical problems). We actually reverse the argument. Instead of agreeing with Grünbaum that we only use Cartesian coordinate systems because they are often more convenient than others, we maintain that we only use non-standard metrics if it is convenient in a certain circumstance. A unique property of the standard metric is its invariance under translation, rotation, and inversion, This is not the case because of some convention, but follows from the homogeneity and isotropy of space. Grünbaum has paid too much attention to the amorphousness of space, which implies the arbitrariness of the unit, and has neglected the symmetry properties inherent to Euclidean geometry.

Grünbaum's remarks could be accepted if they were related to topology, in which, e.g., one does not distinguish between a sphere and an ellipsoid, or a rectangle and a parallellogram. Topology differs from metrical geometry because it lacks a metric. The theorems of topology hold for a figure regardless of how it is deformed in homogeneous strain. Grünbaum, however, directs his conventionalist views to metrical space.

2.9 *The opening up of the spatial modal aspect*

Our critique of Grünbaum's views also pertains to non-Euclidean manifolds although the notion of a general coordinate system does not hold in this case. Non-Euclidean manifolds are in general less symmetric than Euclidean ones. Grünbaum seems to overlook this. Only by tacitly assuming that the said requirement of objectivity (i.e., that the relative position of two subjects be independent of the choice of the reference system) is satisfied is it possible to describe the nature of a non-Euclidean manifold by its metric. This requirement is satisfied in Euclidean space by the rotation, translation, and inversion invariance of its metric. In non-Euclidean space one must either have similar intrinsic symmetries (as in the case of a spherical surface), or refer to some extrinsic instance – for example, to an Euclidean space of higher dimension, or to a rigid body,[43] or to kinematic motion, as is done in relativity theory.

In Gauss' theory of curved manifolds, which shows that the metric

metric with respect to translations along the x-axis, and reflections with respect to the y-axis. His non-standard metric reflects these two symmetries just as well as the standard metric does. But then a semi-plane is not a very interesting example, in particular not for Grünbaum's purposes.

43. Grünbaum B 8ff; Beth D 71.

can be derived without reference to an outside system, it is tacitly assumed that the unit in orthogonal directions and at different positions are the same. Stated otherwise: the metric, and thus the Gaussian curvature depend on the method of measuring lengths adopted on the manifold.[44] Thus one can either start with the symmetries of the manifold, and require that the metric be invariant under the allowed symmetry operations, as is the case for Euclidean or spherical geometry, or start with a rigid definition of length in order to investigate the structure of that manifold. One cannot have it both ways.

We can understand non-Euclidean manifolds in two ways: as an $(n - 1)$-dimensional boundary of an n-dimensional spatial subject (e.g., a spherical surface), or as a manifold whose metric is determined by kinematical or physical laws (as e.g. in relativity theory). In the latter case the homogeneity and isotropy of space are relativized by those non-spatial laws. In the former case they are relativized by the n-dimensional subject whose $(n - 1)$-dimensional boundary functions as a manifold. In both cases the spatial relations between subjects bounded to such a manifold become non-Euclidean because of some restriction – namely, a "boundary condition" in either a literal or an analogical sense. This relativization is characteristic of the opening up of a modal aspect. In kinematics or in physics, we speak of a field as soon as the spatial isotropy and/or homogeneity is lost: a field may either be homogeneous, if it is not isotropic, or it may be neither homogeneous nor isotropic.

Hence we consider Euclidean geometry as having an original spatial meaning, whereas the meaning of non-Euclidean geometry is found by reference either to the numerical modal aspect (in the concept of a boundary), or to the kinematic and the physical aspects.

The spatial modal aspect can also be opened up on the law side by the introduction of multiply connected manifolds. In the simplest case, a linear manifold is open if, for three points, there is one and only one point which lies between the other two. This is the case, for example, with a straight line or a parabola. A linear manifold may also be closed (a circle) or self-intersecting (a lemniscate). Two-dimensional manifolds may be simply connected (e.g., a plane) or multiply connected (e.g., a plane with a hole, a sphere, or a torus). In this case a criterion for being simply connected is given by the concept of contraction. A two-dimensional manifold is called simply connected if any point and any closed curve meet the following two-part criterion: (a) we can uniquely determine whether the point lies

44. Nagel B 244, 246.

inside the curve, and (*b*) (if (*a*) is satisfied) whether the curve can be continuously contracted to the point without leaving the manifold. The surface of a sphere is not simply connected because it fails the first part of the criterion. The surface of a torus does not meet either part of the criterion. In a similar way we can define simply-connectedness for higher-dimensional manifolds, i.e., with the help of the concept of a boundary. We therefore find that these criteria of connectedness have an objective character. We shall restrict our discussion to simply connected manifolds, although multiply connected manifolds are not irrelevant to physics. E.g., gravitational fields and electric fields are simply connected, but the magnetic field around a current bearing conductor is multiply connected. As a consequence, a static electric field can be described by a potential, but a magnetic field cannot.

3. Metric and Measurement

3.1 *Measurement*

In our discussion of the numerical and the spatial modal aspects, both the theory of groups and the concept of isomorphy played an important part. We used the theory of groups as a mathematical theory of relations, and the concept of isomorphy as a mathematical expression of the philosophical concept of analogy. In the present chapter, we want to show that both are very important for the understanding of the mathematical opening up of the physical sciences. Especially since the 17th and 18th centuries, physicists have tried to find numerical and spatial objective descriptions of kinematic and physical relations. The concept of isomorphy enables us to introduce numerical and spatial representations of kinematical or physical states of affairs. The theory of groups provides us with the possibility to give "operational definitions" of the metrics of such representations, and allows us to find mathematical theories for kinematics and physics. Together they form the basis of measurement, and hence of modern empirical science.

Measurement is at the heart of the physical sciences, and therefore it seems justified to devote an entire chapter to its problems. Moreover, it gives us an opportunity of showing the power of the basic distinctions with which we started our investigations even before we apply them to physics and kinematics.

Measurement is the establishment of objective relations between subjects under a law. In science we must find out which modal subject-subject relations can be objectified, and to which law (which metric) that relation is subjected. We also have to discover which structures of individuality are most suited to provide us with standards of measurement. Only then can we discover in which way modal relations are individualized in typical structures.

Measurements are always performed with concrete existing subjects. However, we are mainly interested in universal – i.e., modal, relations and, therefore, we need a theory to provide a bridge between typicality and modality.

The aim of measurement in physics is to obtain an objectification of physical states of affairs – i.e., to represent physical subject-

subject relations by modal numerical or spatial relations. In the former case this means a quantification, in the latter case it can mean a representation in graphs. A direct quantification can be achieved by counting. An indirect one is obtained by measurement, i.e., the comparison of subjects which are comparable because they have an objective property in common.

The possibility of performing experiments and doing measurements is largely responsible for the growth of the physical sciences in modern times. One may wonder why this growth is not present to a greater extent in the social sciences. One reason, of course, is the difficulty of designing relevant experiments because of ethical considerations. There is a second reason, however, which is perhaps more important. It is the lack of a modal metric in the post-physical modal aspects. We shall see that the metric is the indispensable law side of measurement. If our view is correct, the great problems encountered in the social sciences with respect to measurements and their interpretation[1] are largely due to an absence of this metric.

We shall not strive for completeness. Important aspects of measurement, such as the psychological (observational) and cognitive (rational) aspects, will at best be treated superficially. In this chapter we will restrict ourselves to "classical" measurements. Later we intend to show that the so-called measurement problem in quantum physics is not really a measurement problem, because measurements in quantum physics are performed in the same way as described in this chapter. This problem concerns, on the one hand, the application of statistical methods to a number of individual measurements, and, on the other hand, some ontological interpretations of interactions which occur, for example, in measurement processes.

3.2 Comparative properties

According to Carnap and others,[2] the classical distinction of qualitative and quantitative properties is insufficient. There is a third type, called comparative or topological properties. For instance, it is quite meaningless to seek a dichotomy behind linguistic pairs, such as long-short, heavy-light, hot-cold, small-large, fast-slow, old-young, etc. In these examples we should speak of larger than, heavier than, hotter than, etc. We, therefore, speak of a *comparative* attribute if it enables us to put the objects[3] to be compared into a linear order of

---

1. See e.g. Pfanzagl 11: "... measurement in classical physics poses no problems comparable to those in the behavioral sciences ..."
2. Carnap B 8-15; Hempel A 54-58; Stegmüller 17, 27ff.
3. Because we shall not be concerned with measurement as a human act, we shall, from now on, speak of subjects: the objects of measurement are subjects of

more or less. We speak of a *quantitative* distinction if it is comparative, and has a numerical scale. We need to distinguish one more case – viz., if the property is quantitative and its scale is subjected to a metrical law, or, in short, a *metric*. In this case it is not only possible to assign ordering numbers to the subjects which are compared, but also to assign numbers to the subjective relations between them.

These definitions are still not complete. Not only do comparative attributes have an order of before and after, but they also have an order of equivalence. Thus, if we wish to compare any pair of subjects with respect to a comparative attribute, we must either order the two subjects in the form of a more-less statement or show that they are equivalent. For instance, two physically qualified subjects either have the same weight or one is heavier than the other. All physically qualified subjects having the same weight constitute an equivalence class with respect to the property "heavy". Strictly speaking, the subjects are not numerically ordered, but the equivalence classes of their objective properties are.

For a comparative concept we require therefore that there exist an ordering relation, $R_x(A, B)$ – for pairs of subjects A and B – with three possibilities, R_+, R_-, and R_0, subjected to the following rules (cf. Secs. 2.2 and 2.7):

(*a*) One and not more than one of the three possibilities of R_x applies to the pair (A, B): R_+, R_-, and R_0 are alternative and mutually exclusive.

(*b*) $R_+(A, B)$ implies $R_-(B, A)$, and vice versa,
$R_0(A, B)$ implies $R_0(B, A)$.

As a consequence, R_0 applies to the pair (A, A): any subject A is equivalent to itself, with respect to any attribute. R_0 is a symmetrical relation, R_+ and R_- are asymmetrical, and each other's converse. R_+ is called *precedence*, R_0 *equivalence*.

(*c*) For three subjects A, B, and C, each with the same attribute, if $R_x(A, B)$ and $R_x(B, C)$, then $R_x(A, C)$, where R_x stands either for R_+, or for R_-, or for R_0: each relation is transitive.

The three rules (*a*)-(*c*) describe a one-dimensional *quasi-serial ordering* of subjects A, B, C, ... with respect to the attribute R_x. This array is called "quasi-serial",[4] because it is serial except for the fact that several elements may occupy the same place in it. Any attribute R_x that satisfies rules (*a*)-(*c*) will be called a *magnitude*. It serves to objectify the subjects to be compared.[5]

physical and prephysical laws. For the distinction of subjects and objects, see Secs. 1.6 and 2.6.

4. "Linear order" would be a better term, cf. Sec. 2.3.

5. Campbell A, Ch. 6; B, Ch. 1; Hempel A 59; Suppes A 96, 97; Ellis 27; Stegmüller 29ff; Nagel C; Bunge B 36; D 197.

60

Due to precedence, a magnitude always refers back to the numerical modal aspect. But, in the first place, it refers back to the spatial modal aspect provided the magnitude is not a spatial concept itself, such as length. The subjects which are compared are at least equivalent insofar as they share the objective property used for the comparison. Moreover, not the subjects themselves but their equivalence classes are numerically ordered. Finally one can observe that a subject's physical properties, to the extent by which they change, refer back to the kinematical aspect.

3.3 *Scales*

An attribute R_x subjected to rules (a)-(c) can be objectified by assigning numbers to the equivalence classes. If we have practical means of establishing equivalence and the order of the equivalence classes, we call the magnitude measurable, and we can devise a scale, which is thus a numerical objectification of the property concerned. Without any order or equivalence, we can still use numbers to indicate subjects (e.g., football players), but only for identification purposes.[6]

At the present stage, the only restriction applied to such a scale is that the order of the assigned numbers must reflect the serial order of the equivalence classes. Such a scale is by no means unique. A scale (x) can be replaced by any other scale (x') if x' is a monotonic function of x. We may even replace an increasing scale by a decreasing one. A special scale transformation is a linear one:

$$x' = ax + b \text{ (a and b are real numbers, } a \neq 0)$$

If $a > 0$, we speak of a positive linear transformation. If $b = 0$ and $a > 0$, we speak of a dilatation. If $a = 1, b \neq 0$, we speak of a shift. A scale is an interval, ratio, or difference scale if it is unique with respect to positive linear transformations, dilatations, or shifts, respectively.[7]

Consider, for example, Mohs' scale of hardness which ranges from 1 (talc) to 10 (diamond) and is defined by reference to the scratch test. A mineral A is called harder than a mineral B if a sharp point of A scratches a smooth surface of B. A and B are called equally hard if neither scratches the other.[8] Such a scale is called *merely ordinal* because the assignment of numbers to the equivalence classes is completely arbitrary, except for their order. In this respect

6. Nagel C 126; Stevens A 144 speaks of a nominal scale in this case, but I don't consider the word "scale" appropriate because there is no possibility for comparison.

7. Pfanzagl 29; Stevens A 141ff; B 24-26.

8. The scratch test is not strictly transitive; cf. Campbell A 128; B 7; Hempel A 61; there are more reliable tests.

61

it does not differ from, for example, an alphabetical ordering. In particular, the difference in hardness between two minerals designated as 9 and 10 is not related to the difference in hardness between two minerals numbered 4 and 5. Neither does it make sense to say that diamond is twice as hard as a mineral with hardness 5 (apatite).

In contrast to this comparative attribute, consider the metrical attribute "volume". If we compare two vessels of 990 and 1 000 litre with two vessels of 100 and 110 litre, we can meaningfully say that the volume differences are the same in the two cases. The differences are equivalent to the same amount – 10 litre. It is also meaningful to state that a container of 1 000 litre is twice as large as a 500 litre container. Indeed, the scale for volumes is not merely ordinal, but is metrical.[9] The main distinction is that in ordinal scales we can assign numbers to the equivalence classes of the ordered subjects themselves, whereas in a metrical scale it is rather the subject-subject relation which is quantified.

3.4 *The metric*
Most physical scales are subjected to a fourth rule, which is not generally recognized.[10] It extends the meaning of the term $R_x(A, B)$ from a comparative ordering relation to a quantitatively objectifiable subject-subject relation.

All subjects A', A'', ... for which $R_0(A, A')$, $R_0(A, A'')$, ..., form an equivalence class with the subject A. Hence we have equivalence classes $R_0(A)$, $R_0(B)$, ... If $R_+(A, B)$ applies, the equivalence class $R_0(A)$ precedes the equivalence class $R_0(B)$. These equivalence classes form a serial (not a quasi-serial) order. Let us now assume that there exists some non-numerical relation between the equivalence classes $R_0(A)$, $R_0(B)$, ... to be called $R(A, B)$. In some (non-numerical) sense, $R(A, B)$ may be equivalent with $R(C, D)$, and hence there may exist an equivalence class $R(A, B)$. Now we are able to formulate the following rule:

(d) The equivalence classes of the relations $R(A, B)$ for all possible pairs of subjects A and B with respect to some attribute, are elements of a group, isomorphic to a specified group of real numbers.

This isomorphism is called the metric, and a magnitude satisfying rule (d) is called metrical.[11]

9. Nagel C 126, 127; Bunge D 198.
10. A notable exception is Suppes A 265, 266.
11. Bunge D 198.

The relevance of groups to our investigation was already mentioned in Sec. 2.2. We speak of a specified group of real numbers. This specification includes both the interval of allowed numbers, and the group operation connecting them. In many cases the interval is just the set of all real numbers, and the group operation is addition. Then we have a difference scale, such as that for volumes and mass differences. In other cases the interval consists of the positive real numbers, and the group operation is multiplication. Then we have a ratio scale, such as that for volume ratios or mass ratios. According to special relativity theory, the group of all possible relative velocities in one dimension for real moving subjects has an upper bound c, the speed of light. The addition of two relative velocities v and w is given by the group product $(v + w)/(1 + vw/c^2)$, which follows from the properties of the so-called Lorentz group (cf. Ch. 4).

The equivalence classes $R_0(A)$, $R_0(B)$, ... do not form a group with respect to a certain attribute, but the equivalence classes of their relations, $R(A, B)$, do. For instance, if the group operation is isomorphic to addition, negative values must be included, which is not admissible for volumes, whereas it is for volume differences. If the group operation is isomorphic to multiplication one has to take volume ratios as the elements of the group, because the product of two volumes is not a volume, whereas the product of two volume ratios is again a volume ratio.

However, the equivalence classes of the subjects themselves can be considered as relations to the identity element of the group, and can thus be interpreted as a subset of the group. Thus the volume of a subject can be considered as the volume difference with a (fictitious) subject with zero volume, and the set of all volumes is isomorphic to the set of all volume ratios with a subject with unit volume.

Among magnitudes we discern measurable *properties* of subjects, and measurable *relations* between subjects, but the two properties are closely related. Properties also have a relational character, whereas relations have a property-character – compare distance (a relation) with length (a property).

3.5 *Units*
For any metrical attribute, there are three "coordinative principles":[12] the existence of practical means of establishing equivalence and the serial order of the equivalence classes, the metric based on a

---

12. Reichenbach A 135.

group structure, and the arbitrary choice of a unit. The isomorphism between the groups of equivalence classes of non-numerical objective relations $R(A, B)$ and the corresponding group of real numbers does not completely define the numerical values to be assigned to the relations. In Sec. 2.8 we have explained this for spatial magnitudes, arguing from the amorphousness of space. In fact, any metrical attribute lacks an intrinsic metric. The addition group of real numbers is itself isomorphic to the addition group which is generated by multiplying all the members of the first group (x) with an arbitrary real number $c \neq 0$: $x' = cx$ (hence, if $x_1 + x_2 = x_3$, then $cx_1 + cx_2 = cx_3$). For this reason we have to assign the number 1 to some arbitrarily chosen relation $R(A, B)$. The number zero is given to the relation $R(A, A')$ for any pair of equivalent subjects A and A'. With these stipulations, i.e., with the choice of a unit, the metrical scale for addition groups is completely defined (see Sec. 3.8).

Likewise, for groups isomorphic to the multiplication group of real positive numbers (for which if $x_1 x_2 = x_3$, then $x_1{}^c x_2{}^c = x_3{}^c$ for all $c \neq 0$), we assign the number 1 to equivalence relations, and some arbitrary number to some subject A. For example, in the thermodynamic temperature scale, the temperature of the triple point of water is given the number 273.16 K, in order to have a simple relation to the customary centigrade (or Celsius) scale.

Often, metrical scales with a unit refer not only to a group of real numbers, but also to a number *field*, characterized by addition *and* multiplication as group operations. For example, when we speak of a length of 10 cm, we mean a length of 10 times 1 cm. For the addition of length, we need the distributive law for number fields (cf. page 38): 3 cm + 5 cm = 3(1 cm) + 5(1 cm) = (3 + 5)(1 cm) = 8 cm. This is the background of the statement that one can only compare (by adding or subtracting) magnitudes having the same "dimension" (not e.g. 1 cm + 3 cm³) and the same units (e.g., 1 m + 3 cm = 100 cm + 3 cm = 103 cm).

We see that there is still some arbitrariness in metrical scales, but compared to ordinal scales, the arbitrariness is greatly reduced. The use of a scale with a unit is only meaningful for metrical scales. A merely ordinal scale has no unit because, in this case, the assignment of numbers to equivalence classes is completely arbitrary except for their serial order.

3.6 *The theoretical character of the metric*
There are several reasons for stating that the metric as introduced above has a theoretical character. The first reason is that the group structure has the character of a *law*. This means that it is always an (empirically based) theoretical hypothesis to state that a certain

attribute has a group structure. In many cases it is a modal law – i.e., the group structure is independent of the typical structure of the subjects which are objectified in the metric. Only the unit, which is arbitrarily chosen, depends on the typical structure of some subject.

This abstract hypothetical character also comes to the fore because the group always has an infinite number of elements, whereas the number of physically qualified concrete subjects having a certain property may be finite. Thus the metric does not refer to actual but to possible relations.

Furthermore, in actual measurements it is impossible to establish equivalence exactly. We will always have to say that two physically qualified subjects, A and B, are equivalent with respect to a certain attribute within the accuracy obtained by our measuring instruments (including our sense organs). This means that in actual measurements rules (a)-(c) can be violated. For instance, if we have a balance that can only discriminate between masses differing by more than 1 gram, and we have three bodies A, B and C, weighing (according to a more accurate balance) 9.25, 10.0, and 10.75 gram, respectively, then according to our crude balance A has the same weight as B, and B has the same weight as C, whereas, according to the same balance, C is heavier than A. This violates rule (c).[13] On the other hand, the metric describes exact relations among subjects because of its mathematical structure.

It is difficult to give a definition of the notion of accuracy. Starting with Gauss, statistical mathematicians and physicists have developed rules to assign a number to the accuracy with which equivalence can be established. E.g., if we say that the length of a room is (4.21 ± 0.01) metre, we assume that the (in)accuracy of the measurement is 1 cm. The precise meaning of this statement, and how the accuracy can be estimated, are mostly technical matters, and will not be discussed here.[14] It is sufficient to note that in actual measurements we always have a finite accuracy. Theoretically we assume that either $R_+(A, B)$, or $R_-(A, B)$, or $R_0(A, B)$ applies, and that these properties are transitive. In experimental physics $R_0(A, B)$ states: within a certain accuracy, A and B are equivalent with respect to R. But this property is no longer transitive. Thus the rules (a)-(d) form a theoretical, rather than an empirical, basis for the metric, and, in fact, for the whole of physics.

In another respect the above example also shows that the metric is abstract: the metric expressly refers to a group of *real* numbers.

13. Campbell B 30ff; Poincaré B 22; Menger A.
14. See e.g. Campbell B Ch. 9-11; Margenau A Ch. 6; B 163-176; Bunge D 209ff.

But the example indicates that measurements can only yield *rational* numbers, i.e., decimal numbers with a finite number of decimals.[15] Again theoretical considerations allow us to assume that, e.g., length must be assigned real numbers. Theoretical geometry, not experimental geometry, proves the length of the diagonal of a unit square is equal to $\sqrt{2}$. We have seen in Sec. 2.3 that magnitudes as retrocipatory analogies can only refer back to the opened-up numerical modal aspect, i.e., to real numbers or vectors with real number components.

It is sometimes suggested that we only use real numbers for magnitudes because of convenience. Only given real numbers, for example, is it possible to differentiate and integrate functions. However, different metrics are related. Given metrical magnitudes of some kind for a subject, one can calculate metrical magnitudes of another kind for the same subject. Thus, if we know the mass m and the velocity v of a subject, we also know its momentum mv and its kinetic energy $\frac{1}{2}mv^2$. This statement would lose its meaning if the quantities of mass, velocity, energy and momentum did not refer to metrical scales. It also shows that the units for energy and momentum are related to those for mass and velocity. But a superficial inspection of the formulas relating these measurable properties shows that they would also lose their meaning if it were required that they should be represented by rational numbers. Only a real number spectrum can accomplish this.

Finally, we observe that it is not always possible to use the same experimental method to determine equivalence. Extreme operationalists maintain that if we use different methods of measurement, we, in fact, measure different magnitudes.[16] Indeed, it is a matter of theory to connect the results of such measurements.

3.7 *Equivalence, measurement, and the spatial aspect*
The notion of equivalence does not mean that the equivalence classes with respect to every attribute can be ordered in a single linear order. A typical counter-example is the essentially two-dimensional ordering of the equivalence classes of different colours perceived by the human eye.

Also the relative spatial positions of subjects, and forces, can

15. Cf. Campbell B 24; Hempel A 29-39, 67, 68; Grünbaum B 175, 176; Carnap C 88; Stegmüller 58, 90ff; Whiteman 256ff; Bunge C 149; D 207ff; Cassirer 57; Bridgman, in: Henkin 227.
16. Cp. Campbell B 29; Bridgman 10, 23; for a criticism of this view, see Hempel B 123ff; C, Ch. 7; D; Byerly, Lazara.

only be measured if they are first decomposed into their spatial components. In these cases, we have multi-dimensional groups, and we must have recourse to a multi-dimensional metric. In a few cases the metric is complex-numbered, e.g., the impedance in alternating current theory.

In an analogical way the thermodynamic state of a physical system is determined by a set of extensive parameters (cf. Sec. 5.2). Thus it may occur that two systems are partly equivalent (e.g., having the same volume but different energy) or completely equivalent in a physical sense (still having different positions or velocities). All this is possible only because the concept of equivalence itself refers to the spatial order of simultaneity. The numerical order of more and less does not contain equivalence.

All measurements are based on the establishment of equivalence. This means that among measurements two types come to the forefront. First those based on a direct comparison of spatial position (coincidence). Every measuring instrument with a visible scale ultimately depends on this type. We shall return to this type in Sec. 3.10.

The other type depends on a physical analogy with the spatial modal aspect. In Sec. 5.5 we shall see that because of its static character, *force* is a retrocipatory spatial analogy of physical interaction. This type of measurement has two sub-types: measurement based on a balancing of forces (see Sec. 3.8), and measurement based on a thermodynamic equilibrium between a physical subject and a measuring instrument, such as a thermometer (Sec. 3.9). In both sub-types the establishment of equivalence is based on a physical equilibrium state.

Note that velocities and currents can only be measured by their static effects. For instance, an electric current can be measured because it gives rise to a magnetic force.

According to relativity theory, in the opening-process spatial simultaneity is relativized (see Sec. 4.5). This means that if we want to measure attributes of a subject that moves relative to the measuring instrument, we have to take into account this relativizing of simultaneity.

3.8 *Extensive properties: mass*
In this section we consider those relational attributes whose interval of allowed numerical values is the set of all real numbers, with addition as the group operation, so that the number zero corresponds with the equivalence relation $R(A, A')$ for any pair of equivalent subjects A, A'. Now it is also possible to assign real numbers to the subjects themselves. If we call $r(A, B)$ the number corresponding to

the relation $R(A, B)$, and $n(A)$ the number corresponding to the subject A, both with respect to some additive attribute R, then

$$r(A, B) = n(A) - n(B)$$

The set of all possible values $n(A)$ is not necessarily a group, but it is (or is isomorphic to) a sub-set of the group of all possible values $r(A, B)$.

It will be clear that even if (by the choice of a unit) the value $r(A, B)$ is uniquely established, there is some arbitrariness in the value $n(A)$. We can add to it an arbitrary real number (which must be the same for all subjects A). I.e., we are free to choose a zero point for the n-scale without any consequence for the r-scale. In some cases (e.g., length) the zero of the n-scale is obvious. In other cases (e.g., spatial position) the zero of n is completely arbitrary.

We shall now discuss the construction of the metric of an additive or extensive property. We call a property extensive, if it satisfies rules (a)-(d) mentioned above, and if

$$n(AoB) = n(A) + n(B)$$

Here the symbol "AoB" means: "the physical sum of the subjects A and B" – i.e., A and B combined in a physical sense, relevant for the attribute concerned. (Instead of "physical" one can also read "spatial" or "kinematical"). The validity of rule (d) implies that this combination procedure which is isomorphic to the addition of real numbers, leads to a group of relations between the subjects A, B, . . ., isomorphic to the group of real numbers. This combination procedure must be specified in every case. For example, consider the combination of two electrical resistors. If they are connected in series, we add their resistances, but if they are connected in parallel, we have to add their conductance.[17] (Conductivity is the inverse of resistivity). In all cases the addition rule only applies if A and B are disjoint.

Let us suppose that we have a means of determining (within a certain accuracy) whether two subjects belong to the same equivalence class with respect to some extensive property. Then we are able to determine uniquely the number $r(A, B)$ for any two subjects A and B, as is seen in the following example. Suppose we want to compare the masses $m(A)$ and $m(B)$ of the two physical subjects. Our measuring instrument is a balance, which allows us to see whether two subjects have the same mass, and if not, which one is heavier.[18]

17. For a more elaborate discussion, see Helmholtz A; Hempel A 62-69; Menger B; see also Bunge D 200ff.

18. Strictly speaking we compare forces (weights) in a balance. Mass is a numerical analogy of physical interaction, see Ch. 5, and cannot be measured directly. Cp. Jammer C 105 ff.

Now we take p bodies with the same mass $m(A)$ and q bodies with the mass $m(B)$, such that the first collection of p bodies balances the second set of q bodies:

$$| p.m(A) - q.m(B) | < e$$

where e indicates the accuracy of the balance. Accordingly,

$$| m(A) - (q/p).m(B) | < e/p$$

So we find that the mass of A is q/p times the mass of B, within the accuracy e/p. If $m(B)$ happens to be equal to the unit of mass (1 kg) then the mass of A is p/q kg.[19] We observe that this measurement yields a rational number because actual measurements always have a limited accuracy.

Whether a magnitude is extensive or not is not a convention. This fact can be falsified by experiment.[20] It is an empirical fact that mass is an additive property, at least under certain conditions.[21] In relativity physics it is shown that mass is only additive if the added subjects have no relative kinetic energy and no relative potential energy. Therefore, e.g., the mass of a deuteron is less than the sum of the masses of its constituent particles – a proton and a neutron. On the other hand, it is not always the measurement procedure that establishes whether a certain property is extensive or not. There are many extensive properties whose numerical values can only be determined indirectly. Therefore, their metrics depend on other so-called fundamental metrics.[22] A typical example is to be found in thermodynamics in which two key attributes – internal energy and entropy – cannot be measured directly. In fact, a large part of a general course in thermodynamics is required to give proper account of the metrics of energy, entropy, and also temperature (which is not an extensive property).

The distinction between fundamental and derived properties (contrary to measurements) is somewhat arbitrary. For instance, the mass of moderately sized bodies can be measured more or less directly, as indicated above, but the mass of both atoms and stars can only be determined indirectly through the laws of motion.[23]

19. It is more complicated but not essentially different, if we take into account the accuracy with which we can make replicas of A and B. A different but equivalent procedure is described by Campbell B Ch. 2, 3; see also Lenzen 22ff; Suppes A 96ff.

20. Bunge D 199.

21. Mach 268-269.

22. Campbell A 134, 142-144; Hempel A 69, 70; for a critical review of derived measurements and "operational definitions" based on them, see Margenau A. Ch. 12.

23. Cp. Stevens B 23.

3.9 *Intensive properties: temperature*

Sometimes, all properties which are not extensive in the sense defined above are called intensive,[24] but we shall start from the more restricted definition used in thermodynamics. We call an attribute intensive if it satisfies rules (a)-(d), and if

either $n(A \circ B) = n(A) = n(B)$

or $n(A \varphi B)$ is not defined

Thus we have a meaningful interpretation for $n(A \circ B)$ only if $n(B)$ equals $n(A)$. If this is the case, we say that A and B are in equilibrium with respect to the property designated by n.

A typical example is temperature. If we bring two physically qualified subjects into thermal contact, then they will eventually have the same temperature. As long as A and B have different temperatures, it makes no sense to speak of the temperature of their sum. The statement: "If a subject A is in thermal equilibrium with a subject B, and if A is in equilibrium with a third subject C, then B and C are in equilibrium with each other" is sometimes called the "Zeroth Law" of thermodynamics. It is, however, just a part of our rule (c) on page 60, and not only relevant to temperature, but to any equilibrium parameter.[25]

For intensive, as well as extensive, properties the establishment of equivalence is implied in the definition of $n(A \circ B)$: we measure the temperature of a body with a calibrated thermometer as soon as we are confident that the two have the same temperature.[26] However, this method is not sufficient to determine unique relations between bodies which are not equivalent with respect to intensive parameters, as can be done with extensive parameters. Consequently, the scale for an intensive property always depends on the scales for one or more extensive parameters.

Sometimes, this dependence is easily found, as, e.g., the internal pressure of a gas. This intensive property is equal to the force per unit area exerted by the gas on the walls of its container, and force and area are both extensive properties. Thus the calibration of a manometer is in principle a simple matter. For temperature, another well known and important magnitude, the construction of a scale is far more complicated. We shall show this in some detail because temperature is a key concept in physics, and because the following

24. Cf. Hempel A 77, 78; Bunge B 34; D 200; Nagel C 128; Stegmüller 47. For instance, Hempel calls "hardness" an intensive property whereas according to our definitions, it is neither extensive nor intensive. Intensive parameters are also called "potentials".

25. Redlich.

26. Redlich.

discussion is very illuminating for our distinction of merely ordinal scales and modal metrical scales.[27]

The tendency of fluids to expand on heating provides the possibility of measuring temperature by the length of, e.g., a mercury column. The mercury temperature scale is defined such that the temperature of melting ice is given the value 0 °C and boiling water is assigned the value 100 °C. The numerical values for other temperatures are found by linear inter- and extrapolation. In this way the temperature is reduced to the extensive scale for length measurements. We say, e.g., that the temperature is 50 °C, if the height of the mercury column is just halfway between the points for 0 °C and 100 °C.

This merely ordinal scale, though very useful, is rightly called conventional, because it depends on the typical properties of mercury. In fact, any property which depends on temperature could be used instead.[28] If we would take another liquid (like alcohol) we would define the 0 and 100 points in the same way. But now a body having a temperature of 50° according to the mercury thermometer would show a temperature of, let us say, 49° on the alcohol thermometer provided its scale is equally divided between 0 and 100 as is the mercury thermometer. A practical way out of this difficulty is to calibrate the alcohol thermometer against the mercury scale, which makes the alcohol scale non-linear, but this does not make the mercury scale less conventional since it arbitrarily assumes that mercury expands linearly on heating, and that alcohol does not.

In modern axiomatic thermodynamics, temperature is usually introduced as the derivative of energy with respect to entropy, which are both extensive properties (see Sec. 5.2). Given the methods of statistical physics it is then possible to design a temperature scale which is not conventional, except for the choice of the unit. The same scale can also be found by thermodynamic means, and we shall describe this older and rather elaborate method in order to stress its modal universality.[29]

The method starts with a very general principle which can be formulated in two equivalent ways. According to Kelvin, it is impossible that the net result of a cyclical process is such that heat is completely transformed into work. According to Clausius, it is im-

27. For the following discussion, see e.g. Morse Ch. 1-6. Another example of an intensive magnitude is the electrical potential difference. The establishment of its metric between c.1780 and c.1850 caused difficulties similar to those encountered in the development of the temperature scale, cf. Stafleu G.

28. Born 36.

29. Still another method is Carathéodory's; cf. Born 39ff.

possible that the only result of a cyclical process is such that heat is transferred to a warmer body. Both statements are expressions of the physical order of irreversibility (cf. Ch. 6).

The efficiency of a cyclical process is defined as the net work (output minus input) divided by the heat input (discarding the heat lost). If we consider several different cyclical processes, all working between the same temperature extremes T_1 and T_2 ($T_1 > T_2$), then it follows from the principles of Kelvin and Clausius that no cyclical process can have a higher efficiency than a so-called Carnot cycle. The latter consists of two isothermal processes (at constant temperature T_1, respectively T_2), interspersed with adiabatic processes (during which no heat is exchanged and the temperature changes from T_1 to T_2, and vice versa). If the heat input at temperature T_1 is called Q_1, and the heat output at temperature T_2 is called Q_2, then with the help of the conservation law of energy, we find that the efficiency of a Carnot cycle is $(1 - Q_2/Q_1)$. It may be observed that until now we did not need a temperature scale. We only need a means of establishing whether two subjects have the same temperature, and if not, which one is hotter.

Because of their properties, it is clear that two Carnot cycles working between the same temperatures T_1 and T_2 must have the same efficiency, irrespective of the typical structure of the processes involved. Therefore, this efficiency can only be a function of these temperatures, and it is possible to define the temperature scale such that $T_2/T_1 = Q_2/Q_1$. This scale is arbitrarily provided with a unit by stipulating that the temperature of the triple point of water is 273.16 K (for Kelvin).[30]

This theoretical thermodynamic temperature scale (also called "absolute temperature") is – except for the unit – independent of any typical property whatsoever. It is completely of a *modal* character.[31] It is based only on the physical time order as expressed in the Second Law of thermodynamics, and the assumption that heat (i.e., energy flux), an extensive property, can be measured directly or indirectly, which is indeed the case, at least in principle, e.g., via the mechanical heat equivalence. This implies that we can use this modal theoretical magnitude in theoretical formulae. Thus it is only meaningful to state that the mean kinetic energy of molecules in a gas is $(3/2)kT$, if T does not refer to the mercury scale, but to the thermodynamic scale.

Certainly a Carnot cycle is not a practical thermometer. It is the

30. This ensures that the temperature difference between freezing and boiling water at standard pressure is still 100 degrees.
31. Cp. Nagel B 11.

task of thermometry to devise practical thermometers which come as near as possible to the theoretical temperature scale. For instance, by theoretical analysis it can be shown that this scale is identical to one based on the expansion of an ideal gas, which is approximated by dilute gases like helium, argon, and hydrogen, except at very low and very high temperatures. But now the order is reversed. We do not define a scale by using a thermometer with its typical properties, but we use a certain thermometer, according to its own convenience, in a certain situation. Its scale should, as nearly as possible, approximate the modal theoretical thermodynamic scale in order to give results which can be used to corroborate or falsify physical theories. Thus we conclude that *the thermodynamic scale is not based on some convention, but on a theoretic analysis of physical relations.*

3.10 *The spatial and temporal metrics*
We have seen that the measurement of extensive properties like mass and intensive properties like temperature depends on a state of equilibrium between two subjects. In Ch. 5 we shall show that such a state is characterized by an equilibrium between two or more (generalized) forces, and we shall argue that force is a spatial analogy of physical interaction.

At first sight this does not apply to the measurement of length and time. If we want to compare the lengths of two bodies which are spatially remote, we take a metre stick, first measuring the length of one subject and then the length of the second; finally we obtain the difference or the ratio of the two values. But what is our guarantee that the length of our metre stick did not change between the two measurements? Why do we take a solid body as our metre stick and not a rubber string? Is the outlined procedure still valid if the temperature in the environments of the two bodies is not the same? Why do today's physicists take the wave length of a certain spectral line as the fundamental unit of length, and not the length of the standard metre at Sèvres?

Similar questions arise with respect to the measurement of time. By an accurate measurement of time is understood the comparison of a certain time interval with a periodic system: a clock. But how do we know that a certain clock is really accurate, such that it ticks off equal periods? Why do we assume that certain clocks are more accurate than others? Do exactly periodic systems really exist?

Usually one reasons that it is impossible to base the measurement of length on the concept of a rigid body, because this would lead us into a vicious circle: to show that a body is rigid, we need rigid bodies. Similarly, to show that a clock is periodic, we need periodic systems. We shall try to make clear that the real difference is getting into this vicious circle, not getting out of it.

The conventionalist's answer to these problems is more or less as follows. We take a large class of any kind of bodies and compare their lengths. Now under certain circumstances (e.g., equal temperature) a subclass of these bodies have invariant length ratios, whereas other subclasses do not. It is just a matter of convenience to take this subclass as the class of rigid bodies which is used as a basis for the measurement of length. Sometimes criteria of simplicity and fruitfulness are added to this convention. In this framework the question cannot be posed (let alone be answered) *why* the physically qualified bodies of this subclass are more or less equally rigid. As Grünbaum says: "Only the choice of a particular *extrinsic* congruence standard can determine a unique congruence class, the *rigidity* or self-congruence of that standard under transport being *decreed by convention*, and similarly for the periodic devices which are held to be *isochronous* (uniform) clocks."[32]

The answer of today's physicists is quite different. They carry out an analysis of all available physically qualified structures in order to find the most *stable* ones, which are used as standards for measurement. For the criteria of stability, the basic spatial and kinematic laws are presupposed: spatial isotropy and homogeneity, and the uniformity of kinematical time are presupposed in the physicist's choice of the standards of length and time.[33] This is what we mean by saying we have to get *into* the circle. We assume – supported by empirical evidence – that space and time are isotropic, homogeneous, and uniform, we choose a metric that reflects these symmetry properties, we investigate typical structures of individuality to find the most stable ones, we choose a reliable standard according to our requirements, and then we check whether space and time are isotropic, homogeneous and uniform. This is a circle, but not a logical one. It is by no means certain whether such a procedure should inevitably lead to consistent results.

It has been discovered (empirically) that the standard metre at Sèvres, and the diurnal or annual motion of the earth do not give sufficient accuracy if subjected to the criteria of temporal uniformity and spatial homogeneity and isotropy. Therefore, today, the physical units of length and time are based on atomic structures: the typical

32. Grünbaum D 14. Cp. Poincaré A Ch. 2; Stegmüller 18, 35, 86, 98ff. For a critical review of this standpoint, see Nagel B 179ff. Popper A 144, 145 observes that the conventionalist's concept of simplicity is itself conventional, and therefore arbitrary.

33. Margenau A 139; Nagel B 255ff; Lenzen 19.

wave length of a certain spectral line, and the period of another one. These spectral lines are due to electronic transitions, within atoms. Therefore, an absolutely stable system (if it existed), could not be used because no transition would occur in it. But as we shall see (Ch. 10), the stability of a physically qualified system like an atom or a solid is determined by a typical balance of kinetic, potential, and exchange energy, the typicality of which is determined by the potential energy – i.e., by the acting forces. Thus spatial and temporal measurements also rely on a balance of forces, which leads to a typical stable equilibrium state.

All this is not essentially changed in general relativity theory if due account is given to the fact that Euclidean straight lines cannot be determined experimentally, and, therefore, must be replaced by geodesics.[34] If we find that metre sticks do not conform to Euclidean geometry, we can account for this either in a spatial way (assuming non-Euclidean geometry) or in a physical way (assuming a "universal", i.e., modal field of force, like gravitation[35]). This again shows the spatial foundation and the physical qualification of measurement. Thus we find that the "self-congruence" of the standards of measurement are decreed by *consistency* of modal and typical laws, not by convention. If such a presumed consistency between hypothesized modal and typical laws cannot stand up to experimental tests, we have to modify our hypotheses. This is the basis of general relativity theory.[36]

Temporal intervals cannot be measured independently of presupposed spatial laws, and spatial relative positions cannot be determined without dependence on temporal laws.[37] It is impossible to define time and space independently by means of their measurement procedures because in actual physically qualified structures (such as measuring instruments or standards) all physical and prephysical modal aspects are involved.

The metrics for spatial and temporal relations are determined by two modal laws: (1) The uniformity of kinematic time, according to which all subjects move uniformly with respect to each other, insofar as it is possible to abstract from their mutual physical interactions. (2) The transformation laws of spatial and temporal scales, which reflect the fact that there is no preferred reference system (spatial and

34. Mittelstaedt 74. In nearly all physical cosmologies designed so far it is assumed that in the neighbourhood of the earth space-time is approximately flat, i.e., satisfies the pseudo-Euclidean metric of special relativity theory.
35. Nagel B 264; Reichenbach A 26; Beth D 122.
36. Mittelstaedt 87.
37. Whiteman Ch. 5.

temporal *relations*, not spatial positions and temporal moments are relevant). Both traits are found in the classical Newtonian metric and in the metric of special or general relativity, the main difference being that in the latter the spatial and temporal metrics are interrelated, whereas in the former the two are supposed to be mutually independent (cf. Ch. 4).

The conventionalist claims that the Newtonian metric is just as conventional as spatial or temporal scales which are rigidly connected to the typical properties of some individual system. Thus, Grünbaum[38] compares the Newtonian metric for time measurement with the scale based on the diurnal rotation of the earth. Compared with the Newtonian metric this rotation is slightly irregular, and slowing down, because of tidal friction. After an extensive discussion, Grünbaum concludes that ". . . apart from pragmatic considerations, the diurnal description enjoys explanatory parity with the Newtonian one".[39] These "pragmatic considerations" include the fact that in the latter metric the physical and kinematic laws can be more conveniently expressed in mathematical terms. Citing Feigl and Maxwell, he says: ". . . one of the important criteria of descriptive simplicity which greatly restrict the range of "reasonable" conventions is seen to be the *scope* which a convention will allow for mathematically tractable laws".[40]

In this discussion, Grünbaum seems to overlook the group structure of the Newtonian metric, which implies, e.g., that 10 seconds now is as long as 10 seconds tomorrow, in the following sense. Suppose we wish to repeat an experiment in which it is crucial that its duration is 10 seconds. Then we will find (other things being equal) the same result today or tomorrow, or at any other time. This would not be the case, if we would measure time on the diurnal scale (at least if our accuracy is high enough to detect the difference between this scale and the Newtonian metric). The result of nearly every physical experiment would depend on the moment it is done.[41] A conventionalist also rejects the use of such "particular" scale

38. Grünbaum B Ch. 2(A).
39. Grünbaum B 74.
40. Grünbaum B 77; on page 75, Grünbaum admits that ". . . it is a highly fortunate fact and not an a priori truth, that there exists a time metrization *at all* in which *all* accelerations with respect to inertial systems are of dynamic origin, as claimed by Newtonian theory . . ." See also Grünbaum D 59 ff.
41. This is even more striking in the amusing examples given by Hempel A 73, 74, and Stegmüller 73, who discuss a time scale based on the pulse beat of the Dalai Lama or the governing president of the United States, respectively. The outcome of any experiment as described above would depend on the momentary health of these dignitaries. See also Reichenbach A 20, 21, 24.

because it is more convenient to refer to the larger system of the "rest of the universe".[42] But then the definition of the scale is (apart from its epistemological aspects) still a purely subjective matter. For us, the choice of the metric depends on the modal law-subject relation. There happens to be a metric which has a universal character, not because it is applicable "everywhere in the universe", but because it has the character of a natural law.[43]

It is not at all interesting to find scales which depend on the typical individuality of some physically qualified subject. Far more interesting is the possibility of finding a modal metric – i.e., a scale that does not depend on the typical structure of some individual system, and which has a group structure. Only then can *an objective representation of physical states of affairs* be warranted. To declare that all possible non-metrical scales are on a par with modal metrical scales, and that the use of the latter is just a matter of convenience, is a gross depreciation of some of the greatest discoveries in the history of science: the isotropy and homogeneity of space, and the uniformity of time.

The conventionalist's claim is based on the true but irrelevant statement that there are no *logical* grounds for accepting one scale above another one. Reichenbach[44] says: "It is a matter of fact that our world admits of a simple definition of congruence because of the factual relations holding for the behaviour of rigid bodies; but this fact does not deprive the simple definition of its definitional character." However, these "factual relations" are subjected to typical laws, which can be analysed with the help of modal laws, and the "conventional definitions" are based on these laws. It is relevant that there are *physical* grounds for preferring metrical scales to merely ordinal ones.

3.11 *The relevance of the metric to the opening-process in physics*
Physical measurements are either based on a direct comparison with the help of physical scales or on the establishment of a state of equilibrium between two or more forces. Thus we use the force exerted by the earth on any physically qualified subject to measure its mass. A typical instrument is a balance originally designed to compare the masses (or weights) of different bodies. But we can also use this balance to measure other forces. We can compare the force

---

42. Reichenbach A 20, 21.
43. Reichenbach's (A 27) distinction of "universal" and "differential" is erroneously reduced to that between geometry and physics.
44. Reichenbach A 17.

of a spring or an electromagnetic force with gravitational forces. This possibility suggests that "force" is a modal, non-typical concept, such that actual forces, though being of a different typical nature, can balance each other, and thus be measured.

It is one of the aims of science to analyse typical structures in terms of their modal aspects. Therefore, it must be possible to reduce measurements of typical properties of (and typical relations between) concrete things and events to modal relations. Indeed, it is possible to relate measurements of different kinds to each other (we gave several examples above). If this were not the case, we would need a separate scale (including a separate unit) for every kind of measurable property. Scales can only be interrelated if they are subjected to a known metric. In physics theoretical analysis (confirmed by experiment) has shown that all necessary metrical scales can be interrelated such that the number of so-called fundamental units or irreducible metrics is limited to three modal ones (e.g., length, time, and mass) and a small number of units referring to the so-called fundamental interactions. The best known of the latter is the unit of electric charge (or, alternatively, electric current).

Although the number of modal fundamental units in physics is definitely three (probably because there are three non-numerical subject functions of physically qualified typical structures), it is largely a matter of convenience which units to choose as fundamental.[45] Usually these are the units of length (metre), time (second), and mass (kilogram), but, e.g., in theoretical physics one often prefers the set consisting of the speed of light, Planck's constant, and the so-called Bohr-radius of the hydrogen atom.

We have already discussed that the choice of the standards for measurement is based on a modal analysis of typical structures. But the introduction of metrical scales for all relevant properties is itself a necessary prerequisite for a modal analysis of typical structures of individuality. In this analysis the typical relations are analysed into modal subject-subject relations, which can be objectified numerically because they are subjected to a modal metric. A modal metric is also indispensable for theoretical synthesis – i.e., the reconstruction of typical structures of individuality from known modal and general typical laws. In this respect, the modal metric for physical subject-subject relations is just as relevant for quantum physics as it is for classical physics. The success of physics as a so-called exact science testifies to the importance of the metric with respect to measurements on individual systems.

Without the help of a metric it is also possible to objectify magni-

45. Margenau A 235.

tudes. But although numerical values achieved in this way may be convenient in comparing and communicating measurement results, they are useless in a theoretical modal analysis because they can merely describe the order of the subjects, not their modal or typical subjective relations with respect to some attribute.[46]

46. Cp. Campbell A 132-134.

4. Relativity

4.1 *The irreducibility of the kinematical modal aspect*

One of Galileo's greatest contributions to physics, although his own account of it is not quite correct,[1] is the discovery that change of motion – not motion itself – needs a cause. This principle is known as Newton's first law of motion: if no forces act on a body (or if all forces acting on it balance each other), the body will not necessarily be in a state of rest, as was the prevailing view since Aristotle, but it will remain in a state of uniform rectilinear motion.

This statement has been criticized. First, one may observe that forces are defined as causes of change of motion, and therefore it is circular reasoning to say that if there are no forces acting on a body there is no change of motion. Secondly, a state of uniform motion depends on the reference system with respect to which the motion is measured. Once again one may speak of circular reasoning. Now, when introducing fundamental, irreducible concepts, circular reasoning is not always avoidable. As we observed in Sec. 3.10, the problem is not how to get out of the circle, but how to get into it. Irreducible concepts cannot be derived from already known concepts, but have to be distilled from them, to be disentangled from historically grown views which are partly right, partly wrong. It needs giants like Galileo to perform this task.

Moreover, it is not quite correct to say that forces are defined by their static effects or by their effects on motion. This may be called their *modal* determination: in a purely modal sense forces are defined in this way. But we also have to deal with the *typical* manifestations of forces, such as electric, magnetic, gravitational, frictional, and elastic forces. They can be distinguished, although they do not lead to typical motions. We cannot say that a subject under the influence of an electric force "moves electrically", "has an electric acceleration", etc. Nevertheless we are able to recognize these forces in ways other than their action on moving bodies in a purely modal sense. We shall

1. Galileo B 215, 216, 244; Dijksterhuis 383ff, 412; Kuhn A 238; Butterfield 13, 69, 85; Čapek 76; Gillispie 51; Toulmin, Goodfield 223, 224.

defer the discussion of this matter until later. In this chapter we shall concentrate on the second problem mentioned above, the relevance of the reference system. This is possible just because of Galileo's principle. It expresses the mutual irreducibility of the kinematical and the physical modal aspects. Forces and other manifestations of physical interaction belong to the latter.

The relativity of motion implies that it is meaningless to attach motion to a single subject without reference to some other system. This does not mean that it is merely conventional to choose a reference system, even if dynamical effects are not explicitly mentioned. A notable example is Copernicus' heliocentric system of planetary motion, which is generally undervalued by conventionalist authors. It is (erroneously) stated that the replacement of the 83 epicycles of Ptolemy's earth-centered system by 17 epicycles in Copernicus' theory greatly simplified matters, but nothing else since, in principle, both should be considered on a par from a relativistic standpoint.[2]

In fact, the simplification was not really that great[3] since the predictive power of Copernicus' model was not better than that of Ptolemy, and therefore Tycho Brahe's objections against the new theory were sound enough.[4] Copernicus' theory did not win the battle because of its simplicity or quantitative features, but because it proved to be superior in some qualitative features. The assumption that the planets move around the sun, not around the earth, is not merely a change of reference system. It enabled Copernicus to solve several problems:[5] (a) the problem of why Venus and Mercury are

2. Margenau A 96, 97; see also Reichenbach A 211, 217-219. Reichenbach is more cautious than Margenau, but he is mistaken when he prefers the Copernican system to the Ptolemaic one, because the former has a dynamic explanation. Such an explanation is only possible with Kepler's system. Mach 279, 283-284 does not discuss the epicycle theory. He merely states that the Ptolemaic and Copernican modes of view are equally correct. Only the latter is more simple and more practical. "The universe is not *twice* given, with an earth at rest and an earth in motion, but only once, with its *relative* motions alone determinable."

3. Dijksterhuis 325 observes that the introduction of the earth's motion about the sun could not make more than five epicycles superfluous, and Kuhn A 171 states: "Judged on purely practical grounds, Copernicus' new planetary system was a failure; it was neither more accurate nor significantly simpler than its Ptolemaic predecessors . . ." See also Kuhn A 168; Feyerabend B; Gillispie 24-26; Toulmin, Goodfield 175, 179; Koestler 194, 195, 579, 580; Koyré 43.

4. Dijksterhuis 332ff; Kuhn A 200; Feyerabend C 260, 261, F 40ff; Toulmin, Goodfield 184ff; Hanson C 171-249.

5. Copernicus; Dijksterhuis 321-324; Kuhn A 171-180; Hesse A 232; Toulmin, Goodfield 172, 173; Koyré 45ff, 129. The following argument only refers to the annual motion of the earth. Omitting the problem of stellar parallax, we restrict ourselves to some problems relating to planetary motion. See Lakatos B 168-189 for a view similar to ours.

always seen near the sun, and therefore, why these planets' period of revolution in their deferents is just one year; (b) the problem of why Mars and the other superior planets always show retrograde motion when they are in opposition with the sun, and, therefore, why these planets' period of revolution in their epicycles is just one year; (c) the problem of why Venus' appearance is "full" when this planet is far away (small apparent diameter), and "crescent" when it is nearby (large apparent diameter) and showing retrograde motion. The latter argument indicates that Venus moves around the sun, and not in a sphere well below the sun's sphere, as in Ptolemy's model;[6] (d) the problem of the relative distances of the planets: whereas in the Ptolemaic system the distances of the planets to the earth could not be determined, in the Copernican system it is possible to find the distances of the planets to the sun in proportion to the distance of the earth to the sun.

Hence, the Copernican system was accepted because it had greater *explanatory power* than Ptolemy's. This was the case only after Galileo removed the largest objections against the dual motion of the earth, by introducing the ideas of inertia, relativity of motion, and superposition of motion. These objections were concerned with the fact that the motion of the earth had no consequences for the motion of terrestrial objects.

Why Kepler accepted Copernicanism is quite a different matter. Kepler and Galileo moved along parallel tracks. While Galileo removed the said objections, but remained faithful to uniform circular motion, Kepler came to reject the latter. Ptolemy's system can be understood as a marvellous attempt to explain celestial motion in terms of simple uniform circular motion. Uniform circular motion was a kinematic principle of explanation introduced by Plato and maintained by Aristotle and all mediaeval authors, including Copernicus. Eventually Copernicus' system was replaced by Kepler's system, which is nearly the final solution of planetary motion as a kinematical problem.[7] Kepler himself was an arduous adherent to the Pythagorean-Platonic tradition, but since he rated Tycho Brahe's observations higher than any theory, he came to reject circular uniform motion as an irreducible principle of explanation. Because the planets

6. Actually, the fact that Venus' brightness does not vary appreciably during its motion around the sun was used as an argument against the Copernican theory by Osiander in his Preface to Copernicus' work (see Copernicus 22). By pointing to the phases of Venus, Galileo turned the argument in favour of Copernicus' system. See Galileo A 334-339; Feyerabend E 109-111; Kuhn A 222-224.

7. Kuhn A 209ff; Dijksterhuis 335-357; Koestler 227-427; Koyré 117-464; Toulmin, Goodfield 198ff.

turned out to move in elliptical orbits with a varying velocity in his system, Kepler immediately recognized that the theory required a further explanation: not a kinematical one, but a physical one, which was later provided by Newton's theory of gravitation.[8] Newton's theory included linear instead of circular uniform motion as an irreducible principle of explanation.

Newton's first law is sometimes considered to be a special case of his second law (if a force F acts on a body with mass m, its acceleration $a = F/m$). Now, it is argued, if we take $F = 0$, we find that the acceleration is zero, and thus we have derived the first law from the second.[9] However, the second law is only valid if taken with respect to inertial systems. A body on which no unbalanced force is acting moves uniformly with respect to an inertial system. Hence, the first law can be understood as a statement of *existence*, i.e., existence of inertial systems.

This implies the discovery that the physical interaction between two subjects is independent of their common uniform motion with respect to some reference system, just as it is independent of their common position with respect to some spatial co-ordinate system, or the temporal moment at which the interaction occurs. This discovery was already made in classical physics, but plays a far more consequential role in relativity theory. We shall discuss this so-called principle of relativity in the present chapter.

4.2 *Kinematical time*

In the present chapter we shall be mainly concerned with relative motion. This was also the physicist's practice until the beginning of this century. Classical physics was chiefly interested in so-called particle motion – the relative motion of *rigid* bodies. Although kinematic relations were described, kinematic subjects were not recognized. Partial recognition came in the various theories of wave motion, but only with the rise of quantum physics were genuine kinematical subjects (wave packets) considered. We shall defer a study of this problem until Chapter 7, and for the time being only discuss relative motion.

Uniform rectilinear motion is relative. One cannot say that a subject moves, if one does not specify with respect to which other subject it moves. Thus relative, rectilinear, uniform motion is a subject-

8. Kuhn A 153, 245, 252; Toulmin, Goodfield 201.
9. See, e.g., Mach 171-172. This gives rise to the misunderstanding that inertia is due to the mass of the subject. But Newton's first law does not contain any reference to the mass of the subjects concerned. See also Dijksterhuis 519f.

subject relation. On the law side this relative motion presupposes the uniform flow of time as the kinematical time order. To common view it seems rather obvious that time flows uniformly, and that, e.g., an hour today is just as long as an hour tomorrow. As late as the 14th century, however, the day (the time between sunrise and sunset) was rigidly divided into twelve hours, with the effect that an hour in winter was much "shorter" than an hour in summer in northern countries.[10] Clearly this chronology does not allow us to describe kinematical motion as "uniform", and it had to be abandoned with the rise of science during and after the Renaissance.

The idea of time flow is rejected by some philosophers,[11] because there is no motion besides the motion of actual subjects. In our view this argument does not hold, because every law is only meaningful if related to subjects. In the same vein, one may also hold that there is no space, because there are only spatial relations between actual subjects. Indeed, nowadays, most philosophers and physicists agree that there is no space and time in an "absolute", substantial sense. Accordingly, in this book the view is defended that the uniform flow of time is a general, irreducible, modal order of time, as such unbreakably connected to subjective, relative motion.[12]

Relative motion is objectified and measured by the velocity, the ratio of the displacement and the duration of the motion. The displacement and the duration are connected via their common end points. These points are usually called "point events", and if we want to study the objectification of motion, we first direct our attention to them. Generally, an event is something endowed with typical individuality, but in a kinematical, modal sense it has the character of coincidence. The fact that events can be preceded and followed by other events refers back to the serial order of earlier and later. There are events which are simultaneous and this fact refers back to the spatial modal aspect. At first sight it appears possible to order events according to serial and spatial principles in an essentially static pattern of moments, which does not differ in any sense from the quasi-serial order we discussed in Ch. 3.[13]

In order to understand that we must consider a new ordering irreducible to the quasi-serial one, we have to turn to the concept of the modal identity of a modal subject (cp. pages 42 and 54). It is an

10. Whitrow 175.
11. Cf. Part II in Gale.
12. The distinction of uniform time flow as a *law*, and uniform motion as a *subjective* time relation should not be confused with Newton's distinction of absolute and relative time (see Newton 6).
13. Gale 66.

empirical fact that a single identifiable subject can be at different places successively, and that different parts of the self-same subject can also occupy the same place at different moments. This we call motion. It leads to a new ordering, one irreducible to the spatial and the numerical, but which presupposes them. (Attempts to reduce motion to succession in a continuous or dense point set lead to paradoxes like Zeno's). Therefore, point events are not immediate objects in the kinematical modal aspect. As points, they are already objects in the spatial aspect, see Sec. 2.6.

The path of the moving subject (which refers back to the spatial modal aspect), the displacement of the subject, and the duration of its movement are objects in the kinematical aspect. The latter two concepts, displacement and duration, should not be confused with relative position and time difference, respectively. Before these static relations can be used in a kinematical context, they must be opened up to become displacement and temporal duration, respectively. Whereas relative position and time difference relate different subjects, displacement and temporal duration apply to one subject. Displacement and temporal duration are related by the velocity of the movement. The velocity is therefore a numerical objective representation of relative motion. Velocity has a group character in both classical and relativity physics, although the group relations are different in the two theories. This means that point events as common boundaries of the displacement and the duration are second order objects.[14] In particle physics, the path of the motion is usually reduced to the path of the centre of mass of the moving subject, which means that after we objectified kinematic subjects to rigid bodies, we objectify a rigid body to a single point. In field theories this is impossible because the motion of waves is essentially extended.

Hence we find that what is usually called "time" or duration is but an objective relation in the kinematical modal aspect – a relation which gives an objective representation of relative motion. Time receives its serial character because it refers back to the numerical modal aspect, but it is still subjected to the kinematical order of uniformity.

4.3 *Combining velocities*
We have now recognized the difference between two rational numbers as a subjective relation in the numerical modal aspect, and the distance and velocity as objective relations in the spatial and kinema-

14. This is an example of a "complex founded retrociption", cf. Dooyeweerd C 164f.

tical modal aspects, respectively. If in one of these cases we know the relation between two subjects A and B, and the relation between B and a third subject, C, is it possible to find the relation between A and C? In the numerical modal aspect, the answer is yes: if A, B and C are numbers, then $(A - C) = (A - B) + (B - C)$. In multi-dimensional space this simple addition rule is only valid if applied to the co-ordinates of the points A, B, and C. But generally speaking, the distance AC between the points A and C is less than the sum of the distances AB and BC. Thus the addition rule in the spatial modal aspect differs from the one in the numerical modal aspect. What about the addition of velocities?

In classical mechanics the spatial substratum of kinematic motion is an "absolute space" in which distances retain their original geometrical meaning. The time flow, as kinematical order of time, is also considered absolute, and only the numerical time difference or duration on the subject side remains as an objective measure of motion. Accordingly, one may add velocities in the same way as one adds distances in original geometrical space.

As a first approximation this is not too bad. Of course, in many cases original geometrical space will approximate the analogical kinematical space very well. Specifically, this approximation appears to be valid as long as the relative velocities concerned are not too large (i.e., small compared to the speed of light).[15]

However, from experiments such as those of Michelson and Morley in 1887, one has to conclude that the addition of velocities is not valid if the speed of light is involved. Any velocity added to the velocity of light results in the velocity of light itself. Lorentz[16] concluded that distances depend on motion. He tried to explain the phenomenon from the typical structure of matter by reducing the so-called Lorentz contraction of the measuring sticks used in the experiments to an electromagnetic cause. However, in 1905 Einstein showed that this contraction has no dynamical, physical cause, but is entirely of a kinematical nature. But before he could do so, he had to reconsider Newton's concepts of absolute space and time.

4.4 *Einstein's critique of Newton's absolute space*
In classical physics the velocity of some particular moving subject is chosen as a unit, and the time needed to cover the unit of length is the unit of time: time is conceptually measured as a distance. Hence the

15. Analogously, one may add geometrical distances in the same way as numerical differences if the three points concerned are situated on or near a straight line.
16. Lorentz 1-7.

comparison of two movements is reduced to the comparison of two distances covered in the same time. However, the possibility of measuring distances depends on the end points of the distance to be measured. Therefore, Einstein's critique of Newton's kinematics was directed first of all to the use of the concept of spatial simultaneity in kinematics.[17]

If we wish to fix the velocity of a moving subject as the ratio of traversed distance and time difference, we need two clocks to establish the duration of the motion. These clocks, placed at the end points of the covered path, have to be synchronous. How is this established? There is no other possibility but to send a signal from one clock to the other. But then we have to know the velocity of the signal if we wish to determine the time difference between its emission and arrival. To measure this velocity we need two synchronous clocks – and so we are caught in a vicious circle.

Einstein proved there is only one way out of this deadlock. Suppose the signal emitted by clock I at time t_1 is received by clock II at time t', and immediately reflected, returning to clock I at time t_2. We then define the instant $t = \frac{1}{2}(t_1 + t_2)$ on clock I to be simultaneous with time t' on clock II.

This at first sight plausible definition is mistakenly called conventional because the signal is supposed to have the same velocity in both directions – and this presupposition cannot be verified in the above-mentioned procedure of synchronization.[18] Actually it is not really very plausible, and, in fact, even contrary to classical kinematics. If both clocks move (with the same velocity) with respect to a third subject, then the velocity of the signal according to Newton's mechanics is not the same in both directions if measured with respect to this third system. And if we apply this synchronization procedure to two clocks moving with respect to each other, then according to classical mechanics, it is impossible for the signal to have the same velocity in both directions with respect to both clocks. In fact, Newton's absolute space and the absolute, resting electromagnetic ether of the 19th century can be said to be invented to overcome these difficulties. It follows that absolute simultaneity is not valid in analogical kinematical space. Still it is a mistake to call the above-mentioned definition of simultaneity conventional. It is based on the isotropy of kine-

17. Einstein A 35-51; B 52ff. In several respects Poincaré had already formulated the same views before Einstein published his famous paper in 1905. Whittaker B 27 consequently speaks of "the relativity theory of Poincaré and Lorentz", neglecting Einstein's contribution altogether. For a critique of this view, see Holton A 175-179; Grünbaum B 400-409; Whiteman 36, 37.
18. Reichenbach A 123-129; Grünbaum B 342-368, 666-708; D 295-336.

matical space, which does not permit different velocities of light in different directions.[19]

19th-century physics supposed the actual existence of a substantial ether as a physical-spatial substratum of optical and electromagnetic phenomena.[20] In the special theory of relativity Einstein proved that this hypothesis cannot be verified experimentally. Such a "substance" without a psychic object function cannot exist in temporal reality, and it has no meaning to accept its existence merely on metaphysical speculative grounds. Modern physics rightly rejects the concept of an ether.[21]

In an interesting discussion of this matter, the Dutch neo-Thomistic philosopher Hoenen introduces the ether as a "localization medium", wherein distances result immediately, and further points out: ". . . this is not something which follows from perception; no, we have intellectual and immediate insight of it".[22] "Distance" is conceived here in a static-spatial sense. But the concept of distance must be opened up if we wish to treat kinematic subjects. Because no signal propagates with an infinite velocity, an immediately resulting distance has no kinematical meaning. In his criticism of the "neo-positivist principle" (i.e., "That which cannot be measured *in principle* does not exist"[23]) Hoenen discusses the ether theory as well as the simultaneity of spatially separated events.[24] There is, however, an important difference. In classical physics – and in Hoenen's view – the ether is conceived of as an actual existing substance. But simultaneity is a spatial time order, and the thesis of the Philosophy of the Cosmonomic Idea, that all actual existing things must possess an objective-psychical function, does not apply to simultaneity as a scientific conception. With respect to the latter one must not ask about its metaphysical being, but about its modal meaning. In our analysis, simultaneity belongs to the spatial modal aspect, and becomes relativized in the kinematical aspect.

4.5 *The interval as an objective kinematical relation*
Einstein based his theory of relativity on the hypothesis that one singular signal has the same constant velocity (c) with respect to all

19. Bunge B 187-188; Whiteman 36; Sklar 287-294.
20. See Doran; Goldberg; Hesse B; Hirosige; Schaffner; Swenson; Whittaker A, B.
21. Beth C 125ff.
22. Hoenen 152, 304 (my translation). Cp. Van Melsen A 184. Hoenen bases his view on Parmenides' identification of being and intelligibility: "ens est intelligibile", Hoenen 16ff; see also Van Melsen A 54.
23. Hoenen 299ff; Van Melsen A 186.
24. Hoenen 297-306; see also Van Laer.

possible moving systems. It is not necessary that such a signal actually exists. The empirically established fact that the velocity of light satisfies the hypothesis is comparatively irrelevant.[25]

In order to achieve this, Einstein had to amend the Newtonian addition formula for velocities. In the one-dimensional case, two subjects moving with velocities v and w with respect to a third subject, have a relative velocity $(v - w)/(1 - vw/c^2)$, instead of the Newtonian value $(v - w)$. One can easily prove that (a) this relative velocity is independent of the choice of the reference system (i.e., a co-ordinate system with a clock), as it should be; (b) a subject moving with velocity c with respect to one reference system does so with respect to all reference systems; (c) no subject can move with a velocity exceeding the value c with respect to any reference system; (d) this expression approximates the Newtonian one if the velocities are low.

In original space the distance d given on page 51 is independent of the chosen co-ordinate system. Einstein defined the interval s between two point events at positions (x_1, y_1, z_1) and (x_2, y_2, z_2), and at times t_1 and t_2, by

$$s^2 = c^2(t_1 - t_2)^2 - (x_1 - x_2)^2 - (y_1 - y_2)^2 - (z_1 - z_2)^2$$
$$= c^2(t_1 - t_2)^2 - d^2$$

because the velocity of light c must be the same in any reference system. This means, a spherical light wave front emerging from a point source must be spherical in any reference system. This leads immediately to the above "pseudo-Euclidean metric". Einstein demonstrated that this metric of four-dimensional "Minkowski space" is independent of the choice of the moving reference system, i.e., the metric is invariant under all transformations of the Lorentz group. However, the distance d and the time difference $(t_1 - t_2)$ now depend on the motion of the reference system, and can no longer serve as independent objective time relations. They are replaced by the interval which now serves to objectify kinematical subject-subject relations. The interval itself does not describe motions. It is a relation in the opened-up numerical-spatial substratum of the kinematical aspect.

Three cases can be distinguished: s^2 may be negative, positive, or zero. These are topological distinctions. In the first case, $s^2 < 0$, it is always possible to choose a reference system such that (for the two point events under consideration) $t_1 = t_2$. I.e., with respect to that reference system (and all reference systems having the same velocity) the two events occur simultaneously, but at different places. This

25. It seems that Einstein developed his special theory of relativity without having knowledge of the experiments of Michelson and Morley. See Shankland; Bunge B 193; Holton A 261-352. For an opposite view, see Grünbaum B 377-386, 834-837. See also Gutting; Hesse A 246; Swenson; Williamson.

interval is now called "space-like", because it looks like a distance. In other systems of reference, t_1 may be before as well as after t_2. It can be shown that in that case no causal relation between the two events can exist, so that the irreversibility as the physical time order is not violated.[26]

Especially in philosophical treatises it has become usual to call two events "simultaneous" if they have a space-like interval. This use has its merits, but should be avoided. Because we consider the special theory of relativity to be a kinematical theory, we do not accept the definition of "topological simultaneity" as the relation of not being connectable by a physical causal chain or signal.[27] The numerical order of before and after is certainly not meaningless with respect to this kind of event, although its meaning can only be relative – viz., to a third event.[28] Moreover, the relation "topological simultaneous with" is not transitive.

In the second case, $s^2 > 0$, a reference system exists such that the two events occur at the same place ($d = 0$), but at different times. If t_1 occurs before t_2 in this reference system, t_1 occurs before t_2 in every other system of reference. A causal relation between the two events is now possible, and their time sequence is independent of the choice of the reference system. This is not true if we transform our reference system into one in which the time flow is reversed. This transformation (called "time reversal") is kinematically admitted, but should be excluded with respect to physical interactions.[29] In this second case the interval is called "time-like".

In the third, borderline case, $s^2 = 0$, the two events may be connected by a light signal. No reference system exists in which either $t_1 = t_2$ or the two events occur at the same place. But if $t_1 > t_2$, then this is the case in any other reference system (time reversal excluded).

Hence, according to relativity theory, the numerical and spatial aspects of time do not lose their original meaning, but they lose their "absoluteness" when they are opened up by the kinematical modal aspect.[30] This applies to the subject side, where time difference and

26. Bunge A 65ff.

27. Cf. Grünbaum A 410ff; B 28-32, 351; D 22; Reichenbach A 127, 145-147; C 40-41.

28. Bunge B 191.

29. Reichenbach C 42.

30. Hence we reject the view, expressed by Minkowski in 1908: "Henceforth space by itself, and time by itself are doomed to fade away into mere shadows, and only a kind of union of the two will preserve an independent reality", Minkowski 75; for a criticism of the Minkowski formalism, see O'Rahilly 404-419, 732-740.

distance are bound together into the interval, as well as to the law side, where the order of before and after and that of simultaneity become relative to motion. In this respect relativity is a profoundly different than Newton's conception in which the numerical and spatial modal aspects function in closed form with respect to motion.

4.6 The special theory of relativity deals with the kinematical modal aspect

Because the velocity of light c occurs in the metric of kinematical space, one may wonder if this metric refers rather to physical space, or perhaps to electromagnetic wave motion. Both questions should be answered negatively. We offer three arguments for this view, before presenting a more positive argument for the thesis that the special theory of relativity is purely kinematical.

The occurrence of a "typical number" ($c = 3 \times 10^8$ m/sec) in the metric is as such of minor significance. If we measured length and width in centimetres, and height in inches, we would have to define the distance d by

$$d^2 = (x_1 - x_2)^2 + (y_1 - y_2)^2 + (2.5)^2(z_1 - z_2)^2$$

in order to arrive at a consistent geometry. The occurrence of the remarkable number 6.25 in this formula, or that of c in relativity theory, could be avoided by the choice of a coherent system of units. The number c occurs in the metric simply because we retain the use of metres and seconds. The second is a kinematic objective unit, which in principle could be replaced by a unit related to the metre via the metric of special relativity, assuming $c = 1$. This method has practical drawbacks (the speed of light is difficult to measure), but in the formulas of relativity theory, velocities are often given in proportion to the velocity of light, which may thus be considered the natural unit of speed.

In a purely mathematical analysis of the kinematical modal aspect c is the limiting velocity of real moving subjects. A subject moving at higher speed would have imaginary time duration and spatial extension. Such quasi-subjects (recently baptized "tachyons"[31]) may have an abstract, modal meaning, but they will hardly be recognizable as abstractions of real, actual subjects.[32] Even an actual light signal never propagates with the velocity c. This is a limiting velocity which would occur in a vacuum. But a vacuum, although nearly approximated in

31. Feinberg; Reichenbach A 147.
32. It may be called an "axiom of identity" that an identifiable moving subject may be at the same place at different moments, but not at different places at the same moment, see p. 84, 85. Tachyons do not satisfy this axiom.

interstellar space, is itself a limiting abstract concept. No spatial realm is really empty, and in any material medium the velocity of light is less than c. The constant c is the limit rather than the velocity of light. In a medium one may find particles moving faster than light in that medium (this is the phenomenon of Čerenkov radiation), but their velocity is still smaller than c. The so-called phase-velocity of a wave packet may be larger than c, but the phase cannot transmit signals, and the so-called group-velocity of the packet, which is identified with the particle's velocity, is always smaller than c.

The laws of relativity theory have other consequences for a number of physical phenomena which are not necessarily of electromagnetic origin. For instance, all particles having zero rest mass move with the velocity c (in vacuum). This is not only the case with light quanta, but also with neutrinos, and the as yet hypothetical quanta of gravitation. Another consequence of relativity theory is that the measured (objective) mean decay time of moving radioactive particles increases as they move faster with respect to the measuring instrument. This "time dilation" is not only observed in radioactive decay caused by electromagnetic interaction, but also occurs if caused by weak or strong nuclear interaction. The latter cannot be reduced to electromagnetic forces, whereas their velocities of transmission are less than the speed of light. In other words, it might have been possible to discover the laws of relativity theory if one had known only the time dilation of non-electromagnetic phenomena. One could have found these laws if all actually existing signals moved with velocities less than the constant c in the metric.[33] Thus c is not, in the first place, the velocity of light, but rather the velocity of light's propagation in a vacuum is equal to c due to the typical structure of electromagnetic interaction.[34]

Let us now present the positive argument for the thesis that the special theory of relativity is purely kinematical. In Sec. 2.8 we saw that the choice of the unit is arbitrary for spatial coordinate systems in the Euclidean metric. However, in order to be able to give trans-

33. A similar argument is given by Goldstein 200: ". . . the transformation properties must be the same for all forces no matter what their origin. The statement 'a particle is in equilibrium under the influence of two forces' must hold true in all Lorentz systems, which can only be the case if all forces transform in the same manner".

34. For a contrary view, see Bunge B 182ff, who, e.g., defines an inertial frame of reference as one in which Maxwell's equations are satisfied. I think Bunge remains too close to the ". . . historical origins of the theory as far as its leading axioms are concerned" (182), which, of course, are not denied in our treatment. Bunge explicitly states that there would be no basis for the special relativity theory without an electromagnetic field (205).

formation rules between the several possible coordinate systems, we must assume (as is usually tacitly done) that we have the same unit of length in all coordinate systems. In Galilean relativity the same assumption is made. We take for granted that the units of length and of time are the same in all reference systems. This assumption is sufficient to derive the so-called Galilean group of transformations between inertial systems. But the choice of these units, as basic units, should now be scrutinized.

We have seen that time in this context means kinematic time, which is determined by the distance covered by a uniformly moving subject. However, in different frames of reference, we have no right to assume that the unit of length will be the same, and accordingly, we also have no right to assume that the unit of time will be the same. Moreover, it may be questioned whether we ought to take length and time as basic parameters. The parameter which distinguishes the kinematic reference frames is velocity, just as it is distance in the spatial case. Apparently, velocity is a derived quantity, for it is defined as the ratio of covered distance and the corresponding time interval. Because instantaneous velocity can only be approximated in this way, we should be aware that this definition could only be anticipatory. However, kinematical time can only be introduced with the help of a subject moving with constant velocity. Thus it seems appropriate to take velocity as the basic unit in kinematical reference systems, and to demand that the kinematic transformation rules leave the unit of velocity invariant. If we then take this unit to be c, we have the basic hypothesis of Einstein's theory of special relativity. This hypothesis is again sufficient to find the so-called Lorentz-group of transformations between inertial systems. It is an empirical matter to decide whether the Galilean or the Lorentz transformations are valid – there is no logical ground for this decision.

Einstein's paper of 1905, in which he published his relativity theory for the first time[35] was divided into two parts. The first "kinematical part" gives all the relevant formulae of relativity theory, which are applied to the electromagnetic problems in the second part. It is well-known that, even before Einstein, Poincaré came very close to the discovery of the special theory of relativity (see page 87). He denied absolute space and absolute time, referred to a principle of relative motion and a principle of relativity, and sought for invariant forms of physical laws under transformation. "But", says Holton, "the existence of the ether is rarely doubted, for, like Lorentz, Poincaré explained by compensation of effects the apparent validity of absolute

35. Einstein A.

laws in moving inertial systems and maintained the privileged position of the ether".[36] Neither Lorentz, nor Poincaré, took the decisive step made by Einstein – namely to recognize that the relativistic effects have a kinematic origin, rather than a physical one.

4.7 *General theory of relativity*

Both in classical mechanics and in special relativity theory the purely kinematical uniform motion is rectilinear. This is related to the assumption that the spatial substratum is (pseudo-) Euclidean. However, according to Einstein's general theory of relativity, the physically relevant spatial substratum is not Euclidean, but is determined by the temporal and spatial distribution of energy. With respect to this non-Euclidean space, kinematical motion occurs along a so-called geodesic, which is the equivalent of a straight line in Euclidean space. A geodesic is usually the shortest connection between two points (e.g., on a sphere, being a two-dimensional non-Euclidean manifold, a great circle is a geodesic. It may also be the longest connection between two points).

The non-Euclidean character of the metric as determined by the energy distribution affects not only the spatial part of the metric, but also the numerical one. The metric does not only refer to space, but to the whole numerical and spatial substratum of the kinematical aspect, now in its opened-up form. If we define the time flow to be uniform with respect to this reference system, it is no longer uniform with respect to an Euclidean reference system.

If we wish to describe kinematical motion with respect to an Euclidean reference system, we have to explain the fact that this motion is not uniform. We do this by introducing a field of gravitation. We call a (non-Euclidean) reference system in which no gravitation occurs, an inertial system. A non-inertial reference system has accelerated motion with respect to an inertial system. For example, the reference system connected with an artificial earth satellite is an inertial system in which the gravitational field of the earth has been "transformed away". This can only be achieved locally, because it is impossible to find a universal inertial system – i.e., a system with respect to which any gravitational field wherever is transformed away.

We can also do the opposite. If we use a reference system moving non-uniformly with respect to a local inertial system, we experience a gravitational field. We feel an extra weight in an elevator accelerating upwards. A physically more important example is uniform

36. Holton A 187; Poincaré A, B.

rotation. If the earth is considered as a reference system then part of its gravitational field is caused by the earth's rotation and gives rise to centrifugal and Coriolis forces. These forces are sometimes called "fictitious" because they have no dynamic origin, but have a kinematic one. According to Einstein such a fictitious field cannot be distinguished from the gravitational field determined by the spatial energy distribution (principle of equivalence).

This is not completely correct, however, because in contrast to the latter gravitational field, a fictitious field can be transformed away "everywhere". This means that only locally we cannot distinguish between the two fields.[37] Moreover, the fictitious field has no physical source, but has a purely modal origin and meaning, whereas the energy distribution, the source of a real gravitational field, is always connected to some typical structure, by which it can be identified. In extended free falling systems (like the earth in the gravitational field of the sun and the moon) one has detectable *differential* gravitational forces, like those giving rise to the tidal motions of the seas. The existence of a uniform homogeneous gravitational field throughout the universe can be ruled out by symmetry arguments (isotropy of space). Thus we find that real forces (including gravitational force) are expressions of *physical* subject-subject relations, which cannot be said of fictitious forces. In particular, fictitious forces are not subjected to Newton's third law of motion, the law of action and reaction. Because a fictitious force has no physical origin, there is no reaction force.

There were two starting points for the general theory of relativity. (*a*) Newton's theory of gravitation implied immediate action at a distance, and is therefore incompatible with relativity theory.[38] (*b*) Gravitation is a universal interaction,[39] and must also be applicable to systems not endowed with mass, e.g., light signals. The second point is the main reason why we consider the general theory of relativity to be a purely modal theory. All free physical subjects, irrespective of their typical structure, move uniformly in a local inertial system, or they are influenced by the corresponding gravitational field in a non-inertial reference system in exactly the same way. On the other hand, it is impossible to transform away an electro-

37. Bergmann 205f; Bunge B 210ff.
38. Jammer A 171ff; B 257; C 205; Akhieser, Berestetsky 372ff.
39. Newton was the first to realize the universality of gravitation by discovering (*a*) that the force between the sun and the planets, the earth and the moon, and the force causing falling motion, are the same, and (*b*) that the gravitational force on a subject is proportional to its mass, regardless of its typical structure or composition. See Mach 229-234, 241.

magnetic field, which has a typical structure. For example, the influence of this field on the motion of a subject depends on the ratio between its charge and mass, that is, on its internal typical structure. The influence of the gravitational field on the motion of a subject is sometimes said to depend on the ratio of its gravitational and inertial mass. This ratio is equal to one and does not depend on the typical structure of the subject. However, this is merely analogical reasoning, because the notions of gravitational and inertial mass do not apply to light signals which also move along geodesics.[40]

4.8 *The principle of relativity*
Besides uniform linear motion a rigid body is able to perform a uniform rotation without any physical cause. The essential difference is that the latter is possible only with a rigid body (or at least a system whose parts are kept together by an attractive force, such as the planetary system). Uniform linear motion is possible for any subject. In fact, every part of the rotating body experiences centrifugal and Coriolis forces, but due to its internal coherence the force exerted on one part is compensated by that on another part. Therefore, the uniform rotation is a *bounded* uniform motion, not entirely of a general modal nature, since it depends on the typical structure of the body (or some internal force).

In special relativity theory any reference system rotated about a finite angle with respect to an inertial system is itself an inertial system. This has nothing to do with a uniform rotational movement, and applies as well to the coordinate systems in geometrical space. Any coordinate system can be rotated about any angle or displaced any distance without disturbing the relative spatial positions of the subjects (cf. Sec. 2.8). But two different reference systems cannot only be translated any distance or rotated about any angle, they can also move uniformly with respect to each other without generating a fictitious gravitational field which would exist in one system, but not in the other one. On the other hand, in a uniformly *rotating* reference system fictitious gravitational fields must be introduced. This is the case in special *and* general relativity theory, just as in classical mechanics.

Newton knew very well that it is impossible to derive the existence of an absolute system of reference by considering uniform linear motion alone. As a minimum he thought that all rotations could be established experimentally with respect to this absolute space (the famous pail experiment[41]). His main contemporary opponents,

40. Bunge B 207ff; C 400.
41. Newton 10, 11.

Huygens and Leibniz, could not refute his arguments, because their arguments were mainly logical rather than physical.[42] It was not until 1883 that Ernst Mach pointed to a flaw in the reasoning: the experimentally established motion does not occur with respect to an absolute reference system, but with respect to the whole of all matter found in space.[43] Later Einstein corrected this view by showing that the rotation occurs with respect to a local inertial system.[44]

Indeed, the distinction between linear motion, as a purely kinematical motion, and rotation, as a movement which anticipates the physical modal aspect, is only comprehensible if the irreducibility of the kinematic modal aspect is accepted as an empirical fact. Newton as well as Leibniz and Huygens were led astray in their judgment of spatial concepts by the supposed reduction of kinematic relations to spatial relations, or conversely, by the inclusion of geometry in kinematics or mechanics. On the one hand, Newton considered the spatial aspect to be subordinated to the mechanical one.[45] This convinced him of the existence of an absolute space – an idea which, he thought, was confirmed by experiments on rotating systems. On the other hand, Huygens and Leibniz tried in vain to reduce the relativity of motion to the relativity of spatial position. Since spatial relative positions are invariant with respect to both translations and rotations of the coordinate system in Euclidean space, they had to assume that not only linear motions but also circular motions ought to be purely relative in a kinematical sense.[46]

The ideas of Huygens and Leibniz were revived into Mach's principle, which in its original form stated that any kind of inertia is caused by the mutual interaction of matter. It is now generally (though not unanimously) rejected, because it turns out to be very difficult to develop this principle into a satisfactory mathematical theory.[47] The principle of relativity, as formulated by Einstein, is more restricted than Mach's principle.

According to Einstein, the local inertial system serves as a sub-

42. Kuhn B 72; Jammer A 114ff; Reichenbach A 213.
43. Mach 279-286. A critique of Mach's views does not need to agree with Russell B 17, who says that "... the influence attributed to the fixed stars savours of astrology, and is scientifically incredible".
44. Eddington 157-165.
45. Jammer A 93ff.
46. Jammer A 114-124. Even Maxwell mistakenly stated: "Acceleration, like position and velocity, is a relative term, and cannot be interpreted absolutely." Cp. Maxwell B 25, and Larmor's footnotes on this page.
47. Mach 286-290; Reichenbach A 210-218; Graves 298-305; Jammer A 139-141, 190-196; Grünbaum B 418-424; Bunge B 134; Büchel; Mittelstaedt 81ff; Sklar 157-234; Whittaker B 168, 183; Nagel B 203-214.

stratum for physical interactions.[48] Thus it is assumed that light travels along a geodesic just like a free falling massive subject. Especially, Newton's third law, action $= -$ reaction, is supposed to be valid (in a static situation) only if referred to a local inertial system. It has proved to be very difficult to corroborate this statement experimentally, for all available local inertial systems are very much like the pseudo-Euclidean reference systems of special relativity. Usually the latter are taken as the pre-physical substratum of physical interactions.[49]

If we make a transition from one inertial system to another, the interval remains the same. Therefore, the interval is called an invariant. The components $(x_1 - x_2)$, etc., of the interval are changed, however. We call a certain variable (depending on x, y, z, and t) a covariant if it transforms at the transition in the same manner as the four components of the interval.[50] Every mathematical expression or quantity referring to the physical aspect should be either an invariant or a covariant, because of the mutual irreducibility of the kinematical and the physical modal aspects. This is Einstein's principle of relativity, expressed in terms of the Philosophy of the Cosmonomic Idea. It is the same requirement of objectivity we discussed in Sec. 2.8. The electric charge, the internal energy or rest mass, and the entropy of an isolated system are invariants. The total energy plus the momentum, and the electric plus the magnetic field strengths in a point, are covariant variables.

We give an ontological interpretation of the principle of relativity since it is based on the mutual irreducibility of the kinematical and the physical modal aspects.[51] Thus far we have treated the principle of relativity mainly considering the subject side, but it has also bearing on the law side. The formulation of physical laws must be frame-independent, whereas the subjective initial and boundary conditions depend on the choice of frame.[52] In both cases the same thing is meant.

48. Mittelstaedt 78. Unfortunately, Einstein once used the name "ether" for this substratum. This is unfortunate, because the latter has nothing in common with the 19th-century ether.

49. Cp. Whiteman 179.

50. This is somewhat loosely expressed. We shall not bother with the further distinction of covariant and contravariant magnitudes (cf. Landau and Lifchitz B 26).

51. For a contrary view, see Bunge B 183: "The principle of relativity is, in short, (a) a heuristic principle and (b) a metalaw statement – and a normative one not a declarative metanomological statement, for it does not say what is but what ought to be the case."

52. Houtappel et al. 596; Bunge B 86, 87. By the distinction of physical laws from the initial and boundary conditions we meet "Curie's observation" that if the world, in all its details, were invariant with respect to displacement there would be no way to distinguish between two parts. Cf. Houtappel et al. 596.

Because the pre-physical modal aspects are irreducible to the physical one, they can be used to objectify the latter. But in order to make full use of this possibility we have to give due account of that irreducibility. The laws of physics must be independent of time, position, and motion. This implies, e.g., that if we consider different sets of subjects with similar subject-subjects relations we must have similar experimental results. This is the basis of objective experimental research, which must arrive at results, reproducible at any place, at any time, and at any velocity.

In its turn, the frame-invariance of physical interaction gives rise to the conservation laws of energy, linear and angular momentum, and of the motion of the centre of mass, for isolated systems. Hence these laws are related to the irreducibility of the physical modal aspect and the aspects of number, space, and motion. We shall return to this matter in Chapter 5.

5. Interaction

5.1 *Isolated systems*

In Chapters 2 and 4 we have shown subjective time to be a relation between modal subjects, qualified by the modal aspect in question: numerical difference, spatial relative position, kinematic relative motion. To indicate modal subjects – i.e., to abstract from the typical individuality of things, events, etc., in order to study their modal relations – is much more difficult in the kinematical and physical modal aspects than it is in the numerical and spatial aspects.

We shall begin with isolated systems as modal physical subjects. It turns out that *interaction* is the general modal physical subject-subject relation. This means that the possibility of isolating systems is limited. In fact, no pair of physically qualified systems is completely isolated. Belonging to the creation implies interacting with every other created thing. Hence the introduction of isolated systems does not seem to be germane to the problem of time. For, by definition two isolated systems do not interact physically, and so do not maintain a physically qualified subject-subject relation, although they still have pre-physical relations: relative magnitudes, relative positions, and relative motion. On the other hand, if two systems interact we may find it difficult to distinguish them from each other. In classical physics a distinction was made between "matter" and the interaction between "unchangeable material particles", but this distinction is untenable since "matter" itself is, in its physical meaning, an expression of interaction. The interaction between two systems may be so strong that they should be considered as one system.

Once again we meet the problem of the identity of a subject (see pages 42, 54, 84), which cannot be solved in a purely modal way. We shall defer the discussion of the individuality of physically qualified subjects to Chapter 8. But provisionally we can propose a necessary, not sufficient, modal criterion for the identity of a system. We speak of a modal physical subject if it can be isolated.[1] This does not

---

1. Cp. Redlich, who defines an "object" (i.e., a physical subject) as anything that can be isolated, and an "isolated object" as one whose properties remain unchanged whatever changes may happen in its environment.

mean, however, that it loses its individuality as soon as it interacts with another subject, although this may happen. The strength of the interaction will determine whether we can continue to speak of two subjects instead of one. In any case, it appears to be extremely fruitful to speak of the interaction of separate systems, especially if they are isolated except for this particular interaction. In fact, it is a necessary methodological prerequisite for their analysis.[2]

If we want to study the abstract general characteristics of a physical "modal subject", we have to leave aside its typical structure. This means that in the present chapter we shall not discuss the branches of physics which investigate the typical structure of physical subjects: electromagnetism, nuclear and atomic physics, solid state physics, chemistry, etc., and also statistical physics, which studies the behaviour of a large number of interacting systems of a certain kind (presupposing their structural similarity). At present we will restrict ourselves to interactions in which either the "internal" state of the system (thermal physics) or the "external" states (mechanics) are involved in a purely modal way.[3]

The objectification of a physical subject invariably requires use of the concept of a "state". In this concept the identity of the system is presupposed, otherwise it would be meaningless to say that a system can be in different states, or can change its state. Referring to what we discussed above, we find that, strictly speaking, a state can only be ascribed to an isolated system. But, in the same vein as above, we will often be able to speak of the state of an interacting system.

We can distinguish three aspects in the concept of a state. First, the state has a specific numerical value for a certain physical variable. In fact, the state usually has several variables, and is said to be completely determined by a number of variables if all other physical properties of the system can be derived from them. These variables simultaneously determine the state. The number of independent variables necessary to determine the state is the latter's dimension. Secondly, we can consider the state of a system in its spatial relation to other systems if they interact statically – i.e., via a field or a force. Finally, we can speak of the state of motion with respect to some other system. In each case the state is changeable.

Sometimes, the first case is called the internal state and the other

2. Cp. Bunge A 125-134.
3. This distinction of external and internal differs from another one common in mechanics. The forces between parts of a system are considered "internal", whereas forces from outside the system are called "external". In this case the reaction of the system on its environment is not taken into consideration. Cf. Suppes A 294-298, Maxwell B 2.

two are called external states. We shall also use this nomenclature, but we need to remember that the variables which determine the internal state refer as well to other systems: they describe numerical orderings of physically qualified subjects.

Among the modal characteristics of physical subjects the concepts of energy, force and current are the most important. We intend to demonstrate that they always refer back to the numerical, spatial and kinematical relations, respectively. Therefore, their numerical values serve as mathematical objectifications of physical relations. Except for very artificial constructions, physics cannot do without any of these or equivalent concepts, because energy, force and current refer back to mutually irreducible modal aspects. On the other hand, they are strongly related because each is a retrocipation in the same physical aspect. In monistic philosophies, such as were popular in the 19th century, this view is unacceptable. Various attempts to reduce one to the other have always been in vain. In our philosophy it becomes clear why this is impossible.

5.2 *Thermal physics*
The "isolated system" is an abstract concept because no concrete physically qualified subject can be completely isolated from other subjects, and because it does not take into account the individuality structure inherent in any concrete physical system. Nevertheless the isolated system is a meaningful concept, since even in experimental physics walls can be devised which are nearly impermeable to energy and matter transport. However, this concept is especially meaningful as a theoretical concept because it allows us to study modal physical laws.

Thermodynamics (or "thermal physics" as it is now more frequently called) deals with the modal physical properties of macroscopic bodies. It was developed in the first half of the 19th century by Carnot, Mayer, Joule, Helmholtz, Clausius, Kelvin, and others. In the beginning of this century Carathéodory investigated its foundations. However, the axiomatic representation of this branch of physics is still a matter of dispute,[4] so that we shall discuss its hypotheses without pretending rigour or completeness.

Whereas in mechanics a physical subject can often be objectified by a spatial point (the centre of mass), in thermal physics a physical subject is essentially extended: it has connected parts. As a first hypothesis it is stated that any isolated system has a macroscopically unique equilibrium state, designated by a limited set of *extensive*

4. Redlich; Bunge E; Noll.

parameters (cf. Sec. 3.8), such as the volume (V) and the internal energy (U) of the system. Some of these parameters may be determined by the boundary conditions (the volume, or a static field), while others are determined by the internal structure of the system.

To understand the meaning of the extensive parameters, we have to consider some possible interaction, because the state of an isolated system can only have meaning while anticipating some interaction. Suppose two systems A and B are completely isolated. The state of the system AoB consisting of the physical sum of A and B is designated by extensive parameters, whose values are the numerical sum of the values for the separate systems:

$$V(AoB) = V(A) + V(B), \quad U(AoB) = U(A) + U(B)$$

Provided no chemical reaction takes place, the number of moles (N_i) of each chemical species is also an extensive quantity. If a chemical reaction takes place, the number of moles of each atomic component must be incorporated as well.

Now let the two systems interact with each other. According to a second hypothesis, the decrease of any extensive parameter for A equals the increase of the corresponding parameter for system B. During the interaction, the total volume, energy, number of moles (or atoms) of any kind, etc., are unchanged. It seems obvious that if the volume of a system increases, the volume of its surroundings must decrease by the same amount. It is not trivial that this also applies to the energy. The conservation of energy is a numerical expression of the physical subject-subject relation.

The first and second hypotheses imply that the extensive parameters also serve to describe the system when it is not in equilibrium. In this case the description is not unique. Different non-equilibrium states correspond to the same set of extensive parameters. A non-equilibrium state of an isolated system can only be described by the way it was prepared (which may include "waiting a little"[5]). However, according to a third hypothesis, there exists a mathematical function of the extensive parameters, called the *entropy S*, which has a definite value for the equilibrium state of the system, and which can be used to describe the development of a system which is not in equilibrium. If we take the physical sum AoB of two isolated systems A and B, the entropy is the numerical sum of the entropies of each system:

$$S(AoB) = S(A) + S(B)$$

If the two systems interact, the total entropy will stay constant or increase. For two parts of a system, which as a whole is in internal

5. Giles 17.

equilibrium, S is an extensive parameter. But with respect to a system which is not in equilibrium, the increase of S is related to the current between parts of the system, and with respect to such parts (or to different non-interacting systems), S determines the "generalized force" or "potential difference" between them. This hypothesis is called the Second Law of thermodynamics.

Energy is always involved in any possible interaction. This means that energy is a relevant state parameter for any thermodynamic system. An interaction in which energy is not involved, would not lead to equilibrium.[6] This fourth hypothesis is usually formulated in the so-called First Law. It states that the energy increase of any thermodynamic system equals the sum of the work performed on the system, and the heat transferred to it. Here "heat" is the product of the temperature of the body and its entropy increase during the heat transfer, and "work" is related to a change in any extensive state parameter except energy. Work is invariably determined by a change in the boundary conditions. Whether a certain extensive magnitude is relevant to a certain system depends on whether it is possible to perform work on that system by changing that parameter. This means, for example, that the magnetization of a non-magnetic gas is not a relevant state parameter. Thus heat and work are not forms of energy, as is often inaccurately stated, but forms of energy transfer. They are related to currents, and cannot serve as state parameters.

The so-called *intensive parameters* (cf. Sec. 3.9), or, better, *potentials*, like temperature and pressure, can now be introduced in two ways: as partial derivatives of either energy or entropy. In both cases the derivatives are taken with respect to the extensive parameters. In these definitions, the temperature T has an exceptional role, because it is defined in two equivalent ways:
$$T = \delta U/\delta S \quad \text{or} \quad 1/T = \delta S/\delta U$$
Other potentials are defined according to the following alternatives. Let the potential Y or F correspond to the extensive parameter X (X is neither energy nor entropy), then:
$$Y = \delta U/\delta X \quad \text{or} \quad F = -Y/T = \delta S/\delta X$$
The first alternative is the so-called energy-representation, and is older than the second alternative, the so-called entropy-representation, which has advantages for the study of currents (see Sec. 5.6). With this definition of the intensive parameters, the First Law of thermodynamics reads:
$$dU = T dS + \Sigma Y dX \quad \text{or} \quad dS = (1/T)dU + \Sigma F dX$$
The values of the intensive parameters determine the direction of the

6. Callen 44.

interaction. For example, if the temperature of a body A is higher than that of a body B, heat will flow from A to B. If all corresponding intensive parameters are equal for the two interacting systems, A and B are in thermodynamic equilibrium with each other.

In thermostatics (a branch of thermodynamics) the intensive parameters are defined with respect to equilibrium states. In this context it makes no sense to attribute a temperature value to a body which is not in thermal equilibrium. However, if we consider this system as having parts – i.e., as consisting of a large number of small interacting sub-systems, each being near an equilibrium state – we can say that the temperature has different values at different positions within the body. In this case we speak of a temperature field. This is why the intensive parameters are also called "potentials". The spatial gradient of a potential has the character of a force, driving a current. Thus, every extensive parameter determining the state of a system is related to a generalized force and a generalized current.[7] But this is only the case as far as the extensive parameters are related to energy.

If the state of a system is determined by n independent extensive parameters, it can be represented by a point in an n-dimensional "state-space", in which those extensive parameters form the co-ordinates. The coordinate system in this space is not unique. We can replace one or more extensive parameters by intensive parameters (which is useful because intensive parameters are often easier to measure), or by other extensive parameters such as the free energy or the (free) enthalpy, if the so-called Legendre transformations are employed. The latter parameters are often useful for the discussion of special situations, such as the equilibrium state of a system that is not isolated but kept at constant temperature or pressure.[8] The number of dimensions of state-space remains the same in these transformations.

By introducing an additional typical law we can reduce this number. One parameter may be eliminated if we define an $(n - 1)$-dimensional manifold in state-space with the help of a so-called *equation of state* which relates an extensive parameter to its corresponding intensive parameter. Such an equation depends on the temperature and on the typical interaction of the molecules composing the system under study. Thus we can find the ideal gas law, which relates pressure and volume if we assume that the molecules in a gas have no extension and do not interact with each other. For

7. "generalized" because the concepts of force and current are originally defined in mechanics.
8. Morse 96; Goldstein 215-216.

the derivation of the specific heat of the gas we need to know whether the molecules consist of one, two, or more atoms. We can find the Curie law for a magnetic gas (relating the magnetization and the magnetic field strength) if its molecules have non-interacting magnetic moments.

These are clearly limiting ("ideal") cases, but they are still dependent on assumptions concerning the typical structures and (lack of) typical interactions. We can try to account for the extension and the interaction of the molecules in a simplified way, and arrive, e.g., at the Van der Waals equation of state, which even permits one to account qualitatively for the condensation of a gas to a fluid. But the development of the equation of state is not a purely modal matter, in contrast to the framework of thermodynamics as outlined above

5.3 *Conservation laws*

The extensive parameters as used in thermal physics mainly describe the internal state of a system. In this section we shall discuss the relevance of energy for the external state which concerns the system's spatial position and motion. Originally, energy as kinetic and potential energy was only recognized with respect to this external state. The internal state was described by a single variable (mass) for a subject whose extension could be neglected, and by a tensor (moment of inertia) for extended, rigid bodies. The relation between mass and energy was not recognized until 1905.

Mechanics is mainly concerned with the relative motion of material subjects and the simplest interaction, therefore, is one of collision. We speak of a collision between two subjects if the interaction can be assumed to be of short duration, and if one's attention is directed to the consequences of the interaction for the relative motion of the systems. Except for this interaction we may consider the two colliding systems to be isolated, and therefore their total uniform motion, as objectified by the motion of their centre of mass, is uniform before and after the collision, and is not influenced by the interaction.

A collision is called *elastic* if the internal state of both systems is not essentially changed by the interaction. It is called *inelastic* if the state of motion as well as the internal state of at least one system is changed in a physical sense. This internal change itself cannot be described by mechanics, unless it is assumed (as was done in classical physics) that it can always be explained by collisions between particles composing the system. Macroscopically, the concept of an elastic collision is an abstraction. Even the collision of two billiard balls is slightly inelastic. A collision between two molecules is usually called elastic if the collision energy is less than the energy of the lowest

106

excited states of the molecules, but even then the wave packets of the two molecules are reduced (cf. Sec. 6.3) such that their internal state changes. Nevertheless, the concept of an elastic collision is very useful in studying the changing kinematical state of interacting systems.

The external motion of the two interacting systems, considered as a whole and objectified by the motion of the centre of mass, can be described entirely in kinematical terms. This is impossible with respect to the relative motion of the two colliding systems. Their motion must be described in terms of kinetic energy and linear momentum.[9] In the 17th and 18th centuries people quarrelled about the priority of one over the other.[10] In elastic collisions, neither the total kinetic energy nor the total momentum of the two colliding systems are influenced by the interaction. The gain in energy of one system equals the loss in energy of the other, and the same applies to the momentum. This is no longer the case in inelastic collisions. Whereas the motion of the centre of mass and the total momentum are still uninfluenced, the interaction changes the total kinetic energy.

In the 19th century it was discovered that, in this case, kinetic energy is transformed into an other form of energy (e.g., by heating of the colliding systems). It became clear that the total energy (including the internal energy) rather than the external (mechanical) energy must be a constant for a system as a whole.[11] Essentially, this is the content of the First Law of thermal physics. It states that the internal energy can only be changed by an external supply of energy – heat or work. Hence, this law does not say anything about the total energy, but concerns itself with energy differences. It says something about the possible increase or decrease of energy and not about its total value. In the special theory of relativity Einstein showed that the mass of a body is proportional to its internal energy (the famous relation $E = mc^2$) if both are determined with respect to a reference system in which the body rests.

After the unification of these three concepts – mass, internal (thermal) energy, and external (mechanical) energy – it became possible to achieve a clear understanding of the meaning of the con-

9. Only in special cases are energy and linear momentum sufficient. In general, angular momentum, an independent conserved property, should also be taken into account, as was first proved by Euler; cf. Truesdell 239-243, 260.

10. Jammer B 165; Mach 310-314, 360-365; Scott; Cartesians considered momentum or quantity of motion as most important, both in elastic and non-elastic collisions. Leibniz and his followers assumed that atoms were elastic, and hence that vis viva (mv^2) was conserved in atomic collisions. Newtonians assumed that atoms were perfectly hard, such that in atomic collisions vis viva was lost. They derived the conservation laws from Newton's third law.

11. Helmholtz B; Elkana A, B.

servation laws. The constancy of energy and linear and angular momentum of an isolated physical system is dependent on the isotropy and homogeneity of numerical time and space, which are tacitly assumed. By isotropy is meant that there is no preferred direction in space, and by homogeneity is meant that there are no preferred instants of time or spatial positions. Only time and spatial differences count – not some absolute time or position parameter.

It can be shown that the symmetry properties of Euclidean space allow ten "constants of the motion": energy, and the three components each of linear momentum, angular momentum, and the position of the centre of mass. Each is related to some group of transformations under which Euclidean space is invariant. In classical physics these are subgroups of the Galilean group, while in special relativity they are subgroups of the Lorentz group. Energy is related to the homogeneity of numerical time, linear momentum is related to the homogeneity of space, angular momentum is related to the isotropy of space[12], and the centre of mass is related to uniform motion.

This implies that if we find other variables which are constants of the motion, they must be derivable from, identical with, or proportional to the ten constants mentioned above (Noether's theorem[13]). For example, a wave packet's frequency and wave vector are proportional to the energy and momentum, respectively. Planck's constant functions as the universal proportionality constant and therefore has a purely modal character (cf. Chapter 7). In fact, the dependence of the constancy of energy and momentum on the homogeneity of time and space is easier to prove in quantum physics than in classical physics (cf. Sec. 9.5). The relation of the conservation law of energy with the homogeneity of time also implies that this law is restricted if the isolation of the system has a limited duration. This restriction is expressed in the Heisenberg relation, $\Delta E \Delta t > h$. See Sec. 7.6.

In general, the ten mentioned conservation laws are mutually independent. Only in special cases is it possible to relate them. For example, a particle with mass m has the energy E and momentum p related by

$$E = p^2/2m \quad \text{or} \quad E = (m^2c^4 + p^2c^2)^{\frac{1}{2}}$$

---

12. The fact that the conservation law of angular momentum refers back to the spatial isotropy implies that it is not necessarily related to rotational *motion*, as was assumed in classical mechanics. The latter view creates difficulties in understanding the spin of electrons and other elementary particles.

13. This was first proved by E. Noether in 1918; cf. Jammer A 198; Bunge B 49. In classical physics mass (the eleventh variable) is also a conserved quantity.

in classical physics or relativity theory, respectively, and for an extended system consisting of point masses between which only central forces are acting, the conservation laws for angular momentum and for the motion of the centre of mass can be derived from the conservation law of linear momentum.

The conservation law of energy has three aspects: conservation of a numerical amount of interaction; the possibility of transfer of energy from one system to another or from one state to another; and the conversion of one kind of energy into another one. We stress the general, modal character of energy. There are many different kinds of energy, modal (internal, gravitational, kinetic), and typical (electric, magnetic, nuclear). However, they do not stand in isolation. They can be transformed into each other if two subjects interact such that heat or work is exchanged, or potential energy is transformed into kinetic energy. For all these different interactions, the universal modal concept of energy allows us to compare the different kinds of energy with each other, and therefore gives a general objective description of them. This gives energy, as the fundamental numerical analogy of physical interaction, its status of key-concept in physics.

5.4 *Force*
We now want to show that force is a physical concept which refers back to the spatial modal aspect and is entirely of a modal character. It will hardly be necessary to show that force is an expression of a physical subject-subject relation. Newton's third law, "action = − reaction", is usually understood in this sense: the force exerted by a physical subject A on a physical subject B equals the force exerted by B on A, but the two forces act in opposite directions. This also applies to the thermodynamic "generalized forces". Besides forces between spatially remote bodies, we also find forces between the parts of extended bodies (e.g., elasticity).

The static character of force implies that different forces, applied to the same physical subject, can balance each other. This is also possible if the mutually compensating forces are of a different typical nature. For example, an electric force exerted on a charged body can be balanced by the latter's weight. An electric voltage across a metallic wire can be compensated by a temperature difference, preventing an electric current from flowing, which would otherwise be caused by the potential difference (this is the so-called thermoelectric effect). This property of balancing forces of different typical character allows us to measure them, i.e., to compare them with each other (see Chapter 3). At the same time it demonstrates the modal, general character of force.[14] Forces must be added in the same way as spatial vectors. This property depends on the independence of forces acting simultaneously on the same subject and is related to the independence of the spatial dimensions.[15]

The three retrocipatory analogies of physical interaction, energy, force and current, are related to each other. The relation of force and energy can be seen in two ways, namely, via the concepts of work (this section) and of potential energy (Sec. 5.5). If a system on which a force is exerted is displaced in the direction of the force, the latter is said to perform work on that system, which therefore gains energy – e.g., the velocity of the system may increase, because its kinetic energy increases. In mechanics this is expressed in Newton's second law of motion. This is only a particular example of the relation between force and energy. It has no application in thermal physics. Therefore it is unwarranted to define forces with the help of this law[16] (this use explains why we speak of "generalized" forces in thermal physics), although as an operational definition it helps us to define the metric and the unit for force.[17]

The concept of force as related to accelerated motion is probably due to Kepler. It became the cornerstone of Newton's theory (second and third law), and ". . . rose almost to the status of an almighty potentate of totalitarian rule over the phenomena . . ."[18] in its interpretation along the lines of Boscovich,[19] Kant and Spencer. In the 19th century people like Mach[20], Kirchhoff and Hertz[21], who realized the relational character of force, tried to reduce this concept to accelerated motion, whether or not mass was a primitive irreducible concept. They rightly reacted against the many attempts to explain the concept of force (especially gravitational and electromagnetic forces) by appealing to some concealed mechanical action of the ether on moving bodies. In these efforts forces were often treated as substances.[22] We may grant these positivist authors the fact that force is not an irreducible relation, such as space, motion, or inter-

14. The necessity of introducing a modal concept of force was better understood by Mayer and Helmholtz than by Hertz and Mach, see Whiteman 398.
15. Mach 44ff, 242-243.
16. Cf. Nagel B 185ff; Poincaré B Chapter 6.
17. The unit of force is the Newton: $1 \text{ N} = 1 \text{ kg.m/sec}^2$.
18. Jammer B 241.
19. Boscovich was the first to realize that the spatial extension of a physical subject is determined by repelling forces; cf. Jammer B 171ff; Agassi A 80ff; Berkson 25-28; Hesse B 163-166.
20. Mach 302-304.
21. Hertz.
22. Jammer B 224; Suppes A 172, 297, 298.

action. But this spatial analogy of physical interaction cannot be reduced to a kinematic relation.

The nominalistic[23] identification of force with the product of mass and acceleration is objectionable not only because of static effects, but also because force can be specified (electric force, magnetic force, etc.). Neither acceleration nor mass can be specified in this way. It would also be quite meaningless to introduce the concept of "fictitious forces" in accelerated frames of reference (cf. Sec. 4.7), if it were not possible to identify "real forces" as physical subject-subject relations. Therefore, Newton's second law is an *equation* and not an *identification*. It cannot serve as a definition of mass or as a definition of force. In the heyday of classical mechanics, when only the functional, modal character of force was considered, this could be overlooked. But nowadays we are more aware of the mutual irreducibility of special forces and therefore of the asymmetry in the equation $\mathbf{F} = m\mathbf{a}$.[24]

5.5 *Fields*

Another way of relating force and energy conceives of a force as the spatial gradient of potential energy. A force describes the static interaction between two or more spatially remote physical subjects, or within a spatially extended subject. Therefore in many (but not all) cases we can substitute the concept of a field for that of a force. Instead of the force exerted by a subject *A* on another subject *B*, we may consider *A* as the source of a field in which *B* is situated (and conversely). The concept of a field was introduced by Kelvin and Maxwell for electric and magnetic interactions, in order to replace action at a distance by contiguous interaction.[25] A static field enables

23. Dijksterhuis 520.

24. In Newton's formulation of his second law, force is related to a change in linear momentum. We may also relate torque to a change in angular momentum, but this requires an independent law: angular momentum refers to the isotropy of space while linear momentum refers to the homogeneity of space. Torque, as well as force, is therefore a fundamental spatial retrocipation in the physical modal aspect.

25. cf. Agassi A; Berkson; Stafleu H. In fluid mechanics, d'Alembert was the first to introduce the concept of a velocity field; cf. Truesdell 122. Fields were also used to express the forces between the parts of a continuous extended body, e.g., elasticity. The concept of action at a distance, reluctantly introduced by Newton, and criticized by Huygens and Leibniz, is in fact alien to the driving motive of classical physics, which only allowed contiguous interaction between unchanging material particles. Kelvin, Maxwell, and many of their contemporaries tried to save this idea with the help of a mechanical ether. In Sec. 4.4 we saw that the idea of the ether is now abandoned. A substantial substratum for fields is no longer considered necessary.

us to determine the force that the source of the field would exert on another body (a "test body", small enough not to change the field) if it were present at some spatial position relative to the source.

A test body has a "potential energy" in the field. It feels a force equal to the spatial gradient of the potential energy. If the test body moves from one spatial position to another, its gain in potential energy is equal to its loss in kinetic energy (provided we have a simple case of a "conservative" field in which there is no irreversible energy dissipation). Hence for static situations, a field describes a possible (potential) interaction, and a force describes an actual one. A field is a spatial concept, anticipating the kinematical and physical modal aspects, whereas force is a physical concept, referring back to the spatial aspect.

A field becomes actual only when related to currents, e.g., electromagnetic waves. The description of the electromagnetic field by Maxwell's equations, allowed him to give an electromagnetic interpretation of light waves. This showed the possibility of interaction (the exchange of electromagnetic energy) which could not be described in terms of Newton's third law. The meaning of this law is also restricted in the theory of relativity, which is concerned with the relative motion of subjects, whereas forces denote static interactions. Therefore, the concept of force will certainly be relativized when the relative motion of interacting subjects is taken into account.[26]

However, this does not imply a loss of meaning, but rather indicates a deepening of meaning. For example, the electric interaction, which is expressed by Coulomb's law in static cases and involves only static electric forces, becomes electromagnetic interaction as described in Maxwell's equations, which include magnetic "forces". Here we meet the first example of an opened-up force. Although magnetic forces may have many characteristics of forces (e.g., we can balance them by other forces), they lack others (they cannot be considered unequivocally as the spatial gradient of a potential energy). In relativity theory the concepts of electric and magnetic force are united into the concept of an electromagnetic tensor. By a change of kinematic reference frame (a Lorentz transformation) the electric field is transformed into a magnetic one, and vice versa.

Friction is a velocity-dependent force which cannot be considered as the spatial gradient of a potential energy, and cannot be reduced to a field. Friction arises when two systems in contact move with respect to each other, or would do so in the absence of friction.

26. Jammer B 254ff.

It is subjected to the physical time order of irreversibility because it invariably leads to a loss of kinetic energy, which is transformed into thermal energy. But friction has the character of a force, in so far as it can be balanced by other forces. It must be taken into account in the application of Newton's second law of motion. Friction allows interacting systems to move uniformly in situations where they would accelerate in its absence. Also mechanical equilibrium is only possible because of friction. This applies not only to the motion of falling bodies in the earth's atmosphere, but also to all moderate currents in thermal physics. In fact, uniform motion influenced by friction is far more common than uniform motion in the absence of any interaction. The latter is an abstraction showing the mutual irreducibility of the kinematic and physical modal aspects, as was first realized by Galileo (although he expressed it in other terms). His contemporary opponents, who defended the Aristotelian view that every motion needs a cause, could certainly point to a firm empirical basis. Galileo's view, which implicitly recognized friction as a force, was reached because he wanted a consistent description of other phenomena.

However fruitful the concept of a field is, it has its modal limitations, and should not be absolutized. Consider two electrically charged subjects. In classical physics each is placed in the field of the other – i.e., each particle experiences the centrally symmetric field of the other. For a third (test) body, however, the field is that of a dipole which is entirely different from a centrally symmetric one. This problem can be evaded by stipulating that no particle feels its own field, but this can only be maintained for a static field. As soon as we allow the particles to move, we are confronted with the unsolvable problem (a problem both in classical and modern physics) that the particle will feel its own field because the velocity of electromagnetic interaction is not infinite. In quantum field theory the attempt to deduce the structure of the electron from the properties of the electromagnetic field leads to an infinite self-energy for the particle, which can only be eliminated by an unsatisfying trick.[27]

5.6 Current and entropy
The concept of current or flow is a modal physical concept which refers back to the kinematical modal aspect. It is not just a thermodynamic concept because we can also find it in electromagnetic theory, in high energy physics, and in continuum mechanics. Generally speaking, current is a transfer of energy, caused by a generalized force. A heat current is caused by a temperature difference, an elec-

27. Weisskopf 96-128

tric current by an electric potential difference, a molecular current by a gradient in the chemical potential, and a simple water current (a river) by a gradient in the gravitational potential. Very often, the current has a uniform velocity, which means that the driving force is balanced by some kind of friction or resistance, whose strength depends on the velocity of the current.

Current is not merely displacement of energy. In that case we should also speak of a current if a free subject has a uniform motion. However, the latter needs no cause since there is no force or interaction involved. We only speak of a current if it is the retrocipatory kinematical analogy of interaction. Our description shows that "work" is also included in this general concept of current. Just as with energy and force, currents may be purely modal (heat flow, and currents caused by gravitation) or typical (some typical currents are mentioned above). The common, and therefore general, feature of currents is the reference from the modal physical aspect to the kinematic one. Current must be distinguished from accelerated motion, which anticipates the physical modal aspect.

In classical mechanics currents other than work are found only in continuum mechanics. The concept of current depends on the opened-up concept of force, i.e., on a field. The basic equation for the motion of a fluid depends on the conservation of matter which is assumed. This consideration gives the so-called continuum equation which is the starting point for all investigations in this theory. This law is also used in thermodynamics with respect to extensive parameters, and is a direct consequence of the property which gives them the name extensive.

Thus the equation of continuity is not applicable to entropy. As soon as there is some kind of friction or resistance, a current is accompanied by a creation of entropy. Entropy will not increase if there is no current. In the limiting case of the performance of pure mechanical work the increase of entropy is zero, but this is only realized if friction and resistance can be neglected.

One of the reasons we gave in considering energy and force as the most general retrocipations of physical interaction to the numerical and the spatial modal aspects, respectively, is that different forms of energy can be transformed into each other, and that different forces can balance each other. With currents this is more complicated. We speak of the thermoelectric effect if a heat current leads to an electric potential difference, and of the Peltier effect if a temperature difference is caused by an electric current. Hence we find that a certain current J_i can not only be caused by the corresponding generalized potential difference dF_i, but also by other potential differences, dF_j. If these gradients are not too large, the current is proportional to them. Calling the proportionality constant L_{ij}, one finds that

$$J_i = \sum_j L_{ij} dF_j$$

and similar expressions for other currents, J_j. If we thus express the potentials in the so-called entropy-representation (cf. page 104), we have Onsager's relation:

$$L_{ij} = L_{ji}$$

for any pair of currents which can accompany each other.

Although the thermoelectric effect and similar effects were known for a very long time, this general relationship was not discovered until shortly before the Second World War, probably because physicists were used to working with the energy representation in which this relation does not show itself in a simple way.

Currents also play an important role in equilibrium situations – which means that a current is not always caused by a force. For instance, if we have a container with a liquid and its vapour in equilibrium, the currents of vaporization and condensation are not zero, but are equal to each other, such that their total effect is zero. If we consider the fact that vaporization is determined mainly by the temperature in the container and condensation is determined by the number of molecules per unit volume in the vapour, we can understand why there is a strong relation between the temperature and the vapour pressure.

This dynamic description of equilibrium is also extremely fruitful in other parts of physics. For example, application of equilibrium considerations of this kind to electromagnetic radiation by Kirchhoff, Stefan, Boltzmann, Wien, Raleigh, and Planck eventually allowed Planck to formulate the first quantum hypothesis.[28]

We have seen that the creation of entropy is invariably connected with currents. This relation, expressed in the Second Law of thermodynamics, has a purely modal character. But the concept of entropy cannot be grasped completely in a purely modal way. It has a strong relation to the concept of probability (cf. Chapters 6 and 8), and to the idea of a ground state and excited states of a physical system. Both have a typical character.

The mass of a system can be considered (since the acceptance of special relativity theory) as the modal expression of its internal energy. This internal energy is determined by the typical structure of the system, and, as such, by its internal interactions. We have to distinguish the ground state of a system, which is its lowest possible

28. cf. Jammer D Chapter 1.

115

internal equilibrium state, from its excited states, which have higher energy values, and therefore have higher masses. This was not appreciated before the 20th century because the energy differences in chemical excitations are relatively small and do not give rise to measurable increase in the mass of chemical substances. Only in nuclear and subnuclear reactions does the mass of interacting systems change appreciably. This accounts for the approximate validity of the law of conservation of mass in most reactions.

The First and Second Laws of thermodynamics deal only with energy and entropy *differences* and not with total energy or entropy. There is also a Third Law, first formulated by Nernst in 1906, which states that at the absolute zero of temperature any conceivable process would leave the entropy constant (at $T = 0$, $\delta S/\delta X = 0$ for any extensive or intensive parameter X except energy and temperature). This means, e.g., that at $T = 0$ the heat capacity of any system is zero, which is born out by low-temperature experiments.

This can be interpreted in the following way. Given an isolated system at rest, the equilibrium state at $T = 0$ is the ground state of the system, for which the entropy is arbitrarily set at zero. At higher temperatures the state of a system is an excited state, and corresponds to a positive entropy, which will increase with increasing temperature. Thus entropy is an extensive measure and temperature an intensive measure of the amount of excitation of a system. Two systems in thermal contact exchange energy until their rates of excitation, as expressed by their temperatures, are equal. In this way, the concepts of entropy and temperature have also significance for microsystems like atoms and molecules.

At first sight, there is no relation between the concept of a current and the idea of a ground state and excited states. But the latter idea has only meaning if applied to a large number of interacting systems (like the molecules in a gas), such that there is a free exchange of energy between the systems concerned. The distribution of energy over the various excited states comes about in a dynamic state of equilibrium, quite similar to that between a liquid and its vapour.

116

6. Irreversibility

6.1 *The direction of time*

In every post-physical modal aspect we find relations of cause and effect which always refer back to the physical modal aspect. We shall discuss the physical cause-effect relation in Sec. 6.7. For the present it is sufficient to observe that this relation is subjected to the law that no effect can precede its cause. This universal law is the physical time order of irreversibility. We shall argue that (*a*) irreversibility is irreducible to the already discussed temporal orders of before and after, simultaneity, and kinematic flow, (*b*) irreversibility is a universal, modal law, not reducible to laws concerning typical interactions, such as probability laws, and (*c*) as a law irreversibility is correlated to the physical subject-subject relation of interaction.

The asymmetry of time does not occur in the first three modal aspects, as long as they are not opened up by the physical aspect. The numerical order of before and after, the spatial order of simultaneity, and the kinematic flow of time are symmetrical. For instance, a purely kinematical movement is reversible in time – that is, if we reverse the sign of the time parameter in the mathematical description of the state of motion of a subject, we again acquire a possible motion subjected to the same law. But if we consider a concrete moving subject, and do not neglect the physical aspect, we have to take into account friction, which is always present, and a motion with friction is not reversible. Due to friction, every changing system will eventually reach a state of relative equilibrium.

We shall distinguish internal (thermal) and external (mechanical) states of equilibrium. The latter depend on friction. An example would be a ladder resting against a house. The friction between the house and the top of the ladder and that between the ground and the bottom of the ladder supply the necessary forces and torques which maintain the ladder in a state of equilibrium (it does not slip). Internal states of equilibrium, according to thermodynamics, can be characterized by a parameter called entropy. Irreversibility, as the physical time order, is expressed by the so-called Second Law of thermodynamics. Let two systems, which are both initially in internal equilibrium, interact with each other. Then the increase in entropy of one system added to the increase in entropy of the other system is larger than or equal to zero. This formulation is more correct than the often heard expression "the entropy of a closed system cannot decrease", because the entropy of a system which is not in equilibrium is not well defined in thermodynamics. Besides, our formulation makes explicit mention of the correlation of irreversibility and interaction.

The irreversibility of physical time is not merely an addition to the numerical time order of before and after. Whenever we have several interacting systems, we do not have causal chains with a serial order, but causal networks with at best a quasi-serial order (cf. Sec. 3.1).[1] According to Reichenbach this means that if a direction is assigned to one causal chain, which connects two systems in a causal network, a direction is determined for each causal chain in the network. This idea of a network presupposes both the numerical order of before and after and the spatial order of simultaneity. Furthermore, the equilibrium state, the final state in any interaction, has a spatial character because it is characterized by a spatial uniformity of some intensive parameter such as temperature or pressure. Finally the increase of entropy, the irreversible process in which the system approaches equilibrium, reflects a relation between the physical and the kinematical modal aspects.

6.2 *The asymmetry of physical time cannot be reduced to probability*
The irreducibility of the physical order of irreversibility to kinematical motion has not gone unchallenged. A basic motive of classical physics was the reduction of all physical phenomena to motions of particles in a field of force. Physicists attempted, for example, to reduce thermodynamics to statistical physics by explaining the macroscopic laws of the former as the net result of the motions and interactions of molecules, which were assumed to be reversible.[2] Especially Boltzmann is considered to have succeeded in deducing the irreversibility stated in the Second Law from a probability calculation on the motion of the molecules composing a gas. In the kinetic theory of gases Boltzmann showed that the thermodynamic concept of entropy is related to the amount of disorder in the system, and he demonstrated an ordered system to be much less probable than a disordered one. He stated that any closed system will develop from a less to a more probable state, and thought that this explained time asymmetry as a macroscopic statistical phenomenon.

1. Reichenbach C 36 speaks of "lineal order".
2. Nagel B 336-345, C 288-312; Reichenbach C 54ff.

In Chapter 8 we shall offer arguments to support our view that probability concerns the relation between typical laws and individual subjects, and therefore cannot serve as the basis for a modal universal law. At present we observe that (a) the mathematical concept of probability does not involve time asymmetry; (b) therefore, statistical physics has to introduce irreversibility as an independent category in probability theory, (c) this is only possible if irreversibility is correlated to physical interaction, and (d) the alleged reduction of irreversibility to statistical laws is a prejudice.

Boltzmann's derivation of irreversibility, in fact, presupposes temporal asymmetry between the initial and final states.[3] Or rather, it shifts the problem to the question: Why should probability increase in time? One might call this "self-evident"[4], but that is begging the question, because then the asymmetry of time is self-evident. It is not: the asymmetry of time is an irreducible mode of experience, empirically discovered.

Consider a closed system, in internal equilibrium. The entropy is not completely constant, but exhibits spontaneous fluctuations (for instance, Brownian motion). Because of the individual behaviour of the composing particles of the system the latter can only be said to be near equilibrium. Considering a system during such a fluctuation, we can deduce that after a while the entropy will most probably be larger. But we can also show that the entropy (with the same probability) will have been larger some time before. Consequently, it is impossible to deduce time asymmetry from the increasing entropy of a closed system alone.[5]

To meet this objection, Reichenbach developed the theory of "branch systems".[6] If we branch off a system from the universe (that is, if we isolate it physically) its entropy most probably will increase or remain constant. According to Reichenbach this determines the direction of time as that direction in which most thermodynamic processes occur. This is an improvement in so far it makes time asymmetry less dependent on the typical properties of the particular systems we study, but still time asymmetry is introduced beforehand in two ways.

First, Reichenbach only considers those systems which branch off and their *subsequent* development. Secondly, Reichenbach and

3. See, e.g. the discussions on this subject in Gold; see also Landau, Lifshitz A 30; Grünbaum B 240ff; Whitrow 5ff; Penrose 41, 42; Schrödinger 14; Weizsäcker B 233, 240.

4. Zwart 144.

5. Reichenbach C 108ff; Grünbaum B 242; Tolman 146ff.

6. Reichenbach C 117ff; see also Grünbaum B 254ff, E 789 ff.

Boltzmann can prove only that some macrostates of the system are more probable than others. They do not prove that the states of a system are ordered in time according to a monotonically increasing or decreasing probability, such that the direction of physical time can be defined as that of increasing probability. This is a separate statement, not implied in the mathematical concept of probability.[7] As a separate law it is yet another expression of the asymmetry of time.

In statistical calculations one also has to correlate this asymmetry with physical interactions, as can be verified in modern treatments of thermal physics. For instance, Kittel[8] introduces the concepts of entropy and temperature with the help of a simple system: a linear chain of spins. For the sake of calculating the entropy etc., one assumes that the spins do not interact with each other. Then it is possible to calculate which macrostates are more probable than others. But, in order to show that a particular state will change such that its probability increases, one has to assume that the spins do interact. Thus the temporal irreversibility cannot be obtained without explicit reference to physical interactions. Therefore, we may conclude that, even in statistical physics, irreversibility is an independent category, correlated to physical interaction. Irreversibility is an irreducible law, and therefore need not be introduced into physics only via probability laws.

It is a widespread misunderstanding that irreversibility is necessarily connected to probability laws. For instance, Lindsay and Margenau distinguish "dynamical laws" from "statistical laws" by two criteria: (a) dynamical laws are deterministic and operate with absolute certainty, whereas statistical laws are only capable of establishing probabilities (they hold for a great number of individuals and lose their meaning if applied to a small number of them); (b) dynamical laws describe reversible processes and statistical laws deal with irreversible phenomena.[9]

However, the first criterion applies to the distinction of modal and typical laws while the second assumes the mutual irreducibility of the kinematical and physical modal aspects. Lindsay and Margenau overlook the fact that any concrete process can be described statistically, and has both reversible and irreversible aspects. Thus the law which describes the motion of a falling body is called dynamical. According to criterion (a) this implies determinism, which can only be maintained if the falling body is not too small, i.e., if Brownian motion and quantum effects can be neglected. And according to

7. cp. Reichenbach C 143.
8. Kittel Chapter 4.
9. Lindsay, Margenau 201.

120

criterion (*b*) this implies reversibility, which could only be true if friction did not exist. On the other hand, heat conduction is supposed to be governed by a statistical law, although if one makes the same approximations as one did with falling bodies the law may be called deterministic. Heat conduction also involves some reversible aspects. For example, the conductivity is independent of the direction of the current in homogeneous media.

The background of Lindsay and Margenau's distinction is that every time a physicist finds a reversible phenomenon he looks for a deterministic interpretation, whereas as soon as he finds irreversibility, he tries to explain it statistically. But this is just narrowmindedness. The proper way to compare these laws is, either to abstract from concrete reality in order to study merely modal laws, or to study the typical individuality structure of the physically qualified subjects constituting a macroscopic body. The distinctions deterministic (universal) – statistical (typical), and reversible – irreversible are not parallel as Lindsay and Margenau suggest, but orthogonal (cf. Sec. 1.2).

6.3 *Irreversibility also applies to microprocesses*
The assumption that the reduction of thermal physics to statistical physics implied the explanation of irreversibility on a macroscopic scale was based on the hypothesis that the interactions between molecules are completely reversible. Since the rise of quantum physics it is clear that irreversible processes also occur on a microscopic scale. This is very obvious with respect to the spontaneous processes which occur, for example, in radioactive nuclei or activated atoms and molecules, and which always involve the transition from a high energy level to a lower one. In 1917 Einstein discovered that these spontaneous processes must be distinguished from stimulated ones, which are in a sense reversible (see page 123).

But even the motion of, e.g., an electron can no longer be considered completely reversible. Its relative place and motion are represented by a so-called wave packet, which is the sum of a number of infinitely extended waves with different wave lengths and amplitudes and mutual phase relations such that the total amplitude of the waves is appreciable only within the wave packet. Outside the packet the composing waves have a resultant zero amplitude. In an interaction such as the collision between the electron and an atom, the electron's wave packet is reduced to a relatively small size, and after the collision this reduced wave packet will gradually extend.

This expansion is irreversible. From a kinematical point of view, it is quite conceivable that one could obtain a contracting wave packet by reversing the motions of all composing waves (preserving their phase relations), but physically no interaction can be designed for this construction. The production of a wave packet and its *subsequent* development (which already presupposes time asymmetry) need an explanation which cannot be given in kinematical terms only. They need a physical explanation, irreducible to a kinematical one. The reduction of the wave packet occurs in any interaction of microsystems, not only in interactions of a microsystem with a macrosystem, e.g., in a measuring process. The latter is often suggested by adherents of the so-called Copenhagen interpretation of quantum mechanics.[10] In our opinion, the micro-macrosystem interaction characteristic of the measuring process is merely a special case of physical interaction.

This is not contrary to the existence of the so-called principle of detailed balance in equilibrium,[11] which is closely related to the principle of reciprocity[12] or micro-reversibility.[13] No one doubts the validity of some principle of overall balance as a necessary (though not sufficient) condition for equilibrium, and its value as a guiding principle for research. But often it is interpreted uncritically as implying the time reversibility of any microprocess. This interpretation overlooks the fact that spontaneous processes (for example, Brownian motion) occur in a state of equilibrium. Also, interactions which result in the expansion of wave packets can be compensated by new interactions without having this process become reversible. Mutually compensating processes certainly do not need to be each other's time reverse. They may be completely different processes, provided they compensate each other's effects.[14] And even if the processes are reversed with respect to their typical structure, they are not necessarily each other's temporal reverse. It suffices that they are equally probable.

In most cases the principle of detailed balance cannot be proved from "first principles". In some cases (namely, those involving magnetic interactions) one can prove that time reversal is not assumed by the principle of detailed balance.[15] In other cases physicists claim that it is. For instance, in classical physics an elastic collision is said to be reversible in time. (In fact, this is only the case if the interacting force is spherically symmetric). This means that, if one would reverse

10. Grünbaum B 249; Ludwig 181.
11. Tolman 161ff, 521; Kaempffer Chapter 28.
12. Kaempffer 255.
13. Messiah 673; Tolman 163.
14. cp. Tolman 114ff, 162: "... in general ... processes which are the inverse of each other do not exist ..."
15. Kaempffer 258-261; Messiah 675.

the time parameter (meaning the reversal of all velocities) in a certain state created after the collision, the collision process then proceeds in the reversed time direction, and one returns to the initial state (with reversed velocities).

In quantum mechanics time reversal with respect to a collision means that if an initial state a corresponds with a state b with a certain probability, then an initial state b' corresponds to a final state a' with the same probability (the primed state is equal to the unprimed state, but with reversed velocities or wave vectors). Although this is a necessary condition for equilibrium in a gas (for example, if the equilibrium arises by means of elastic collisions among molecules) a statement of this kind is not valid with respect to spontaneous processes.

For instance, in Einstein's derivation of Planck's distribution law for black body radiation, he established that the probability for stimulated emission is the same as for stimulated absorption of radiation. But he could only arrive at Planck's formula when he added a third mechanism, the spontaneous emission of radiation, which has no reverse.[16]

Both in classical and in quantum physics it is time in its kinematical aspect which is reversible in those cases where the principle of detailed balance is applicable. In fact the principle only states that there are processes which are symmetrical with respect to kinematical time, just as there are atomic and molecular structures which display spatial symmetries. But, as soon as we study the process of interaction itself, irreversibility is unmistakably present.

6.4 *Time asymmetry concerns subjects, but is a law*

Whoever does not recognize the modal distinction between the kinematical and physical aspects must feel forced to deduce the factual irreversible behaviour of physical systems (including friction) from the statistical result of a large number of reversible processes, that is, finding reversibility on the subject side.

This is also the case in Reichenbach's theory of branch systems. Even if one assumes that the macrostates of a system are ordered according to a monotonically changing probability, one has to prove that the direction of this change is the same for all physical systems. By relating the branch systems to the universe Reichenbach thinks he is able to determine a universal direction of time, which is not a law, but a factual subjective property of the universe. Thus he has to assume that the universe as a whole has an increasing entropy, and

16. cf. Jammer D 112-114.

discusses the possibility that this entropy will decrease at some time.[17]

Meanwhile, the neo-positivist Reichenbach realizes that such considerations are rather speculative, for it is hardly possible to say anything meaningful about the physical properties of the universe (total energy, total entropy),[18] no more than about its relative motion, or spatial position, or even about its number (is the universe a member of a *class* of universes, eventually the only member?). It is very doubtful whether the universe can be treated as a subject with properties like other subjects (Reichenbach states, e.g., that the universe is a closed system, without explaining what this should mean).[19]

Reichenbach's theory is invalid if it is not assumed that the present state of the universe is a state of low entropy. There are at least three objections to this assumption. First, the entropy of a system can only be defined if the system is isolated and finite, and even if we pretend to know what this means with respect to the universe, we do not know whether it is factually the case. Second, the entropy of a system is only defined if the system is in a state of internal equilibrium, and Reichenbach's theory assumes that this does not apply to the universe. Third, Reichenbach relates entropy to probability, but his notion of probability ". . . is always assumed to mean the limit of a relative frequency"[20] – therefore, it does not apply to a single system, such as . . . the universe. What remains of the initial assumption is that the universe is not in a state of equilibrium. We fail to see how one can draw any conclusion from this negative statement.

The universality of physical time asymmetry does not rest on its relation to the universe, but on its being a modal *law*, having universal *validity*. This law does not depend on the empirical fact that all physical processes, which have been observed till now, turn out to develop in one direction of time. We would rather say that the possibility of physical processes depends on the irreducible physical time order which is a basic modal law.

According to Grünbaum, ". . . the complete time symmetry of the basic laws like those of dynamics or electromagnetism is entirely compatible with the existence of contingent irreversibility".[21] But he

17. Reichenbach C 117ff. Reichenbach's reference to the "universe" is criticized by Grünbaum B 261ff. See also Dooyeweerd D 629ff; Popper A 196ff calls explanations which depend on a particular improbable state of the universe "speculative metaphysics", see also Popper F. The attribution of subjective properties like temporal duration and spatial dimension to the universe leads to antinomies, as was first discovered by Kant (A 420-433, B 448-461).
18. Reichenbach C 132, 133.
19. Reichenbach C 135.
20. Reichenbach C 123.
21. Grünbaum B 277

admits that it is rather meaningless to call the laws of motion "basic" and the law of irreversibility a "universally valid" statement about "contingent facts".[22] This artificial construction is invented to reconcile reversibility with irreversibility, by relating the former to laws and the latter to facts. It is very strange indeed, that, e.g., the reversible laws of electromagnetic wave propagation are called "laws", *although* they are only valid in a very limited case (namely, in the absence of any absorbing physical subject), whereas the irreversible law of entropy increase is denied the status of a law, *because* it is only valid in a very limited case (namely in the absence of spontaneous fluctuations). This strange attitude can only be understood if we keep in mind the basic motive of classical physics.

In contrast to this outmoded view, we relate reversibility and irreversibility to mutually irreducible modal aspects of temporal reality. Just as static stimultaneity is retained in the kinematical modal aspect as the borderline case of a state of rest, so we can often find reversibility as a boundary case in the physical modal aspect – for example, electromagnetic wave motion in a vacuum or the motion of bodies when friction can be neglected.

In Sec. 6.2 we concluded that the distinction of reversibility and irreversibility is not parallel but orthogonal to the distinction of typical and modal laws. Now we find that it is also orthogonal to the distinction of law and subject. Therefore, any attempt to reduce one to the other is bound to fail.

6.5 *Interactions*

We agree with one point in Reichenbach's theory of branch systems – namely, the significance of the interaction in our problem – although we do not like to speak about the interaction of a system with the universe.[23] After an interaction with another system, the state of a physical subject will change gradually, until after some time it reaches a state of equilibrium. Thus the first effect of a physical interaction is to disturb the pre-existing equilibrium states of the interacting systems. After the interaction the system approaches a new equilibrium state, i.e., a state of uniform temperature, pressure, chemical composition, electrical potential, etc. This development is irreversible, that is, no system in equilibrium will spontaneously move out of equilibrium, anticipating a future interaction with another system.

---

22. Grünbaum B 273. Both Popper and Grünbaum have pointed out that there are physical processes whose irreversibility cannot be reduced to "entropic" irreversibility; see Grünbaum C, E; Popper F.
23. Reichenbach C 117; see also Grünbaum C.

I have the impression that Reichenbach overlooks this difference. He seems to assume that the entropy of each branch system increases steadily until it is reunited with the universe.[24] In fact, the entropy of any branch system becomes constant after a while, remaining so until it contacts the universe again. Then the entropy will change, but only after this event, not before.

The error in Reichenbach's theory is not simply that he begins with the analysis of a single subject (he admits that to be impossible)[25], but that he does not break *radically* with this method. He tries to deduce the time direction of the universe, as a collection of single systems, from the fact that each system alone tends to a state of internal equilibrium. In our view, the analysis of modal physical time has to start with the *relations* between subjects, especially between pairs of subjects, just as is the case with numerical, spatial and kinematical time.

Of course, if we begin by stating that in any two interacting systems the entropy never decreases, we can extrapolate this fact to the "universe", provided we understand the latter to be as many interacting systems as we like. Put this way, the statement that the entropy of the universe will always increase is valid though quite useless, whether or not this "universe" has a finite limit. This is not the case in Reichenbach's theory which is only applicable to a finite universe. Moreover, we see that this notion of the "universe" cannot be used in Reichenbach's theory, because it presupposes what he wants to prove. Our theory is purely relational, whereas Reichenbach's needs an "absolute" universe.

From the fact that a closed system is not in a state of equilibrium we cannot deduce with certainty that it has interacted with another system some time before: it may be a spontaneous fluctuation. On the other hand, interaction always disturbs equilibrium. Therefore, interaction, rather than the approach to equilibrium, is an irreducible expression of modal subjective physical time.

6.6 *Initial and boundary conditions*

A physical system cannot be described without taking into account its interaction with other systems. Often this interaction can be contained in the so-called boundary conditions. This is the only acceptable interpretation of "the universe": the environment of the system under consideration, i.e., the spatially continuous representation of

24. Especially figure 21, on page 127 of his book, is in my opinion a very inadequate and probably misleading representation of Reichenbach's own views, and certainly of what actually happens. See also Grünbaum E 789, 794, 795.
25. Reichenbach C 117.

the physical relation of the subject with all other subjects. The simplest boundary condition is a rigid wall, which must be understood in a physical rather than a spatial sense. It is an infinitely high and infinitely steep potential energy. Furthermore we can distinguish thermally conducting walls, movable walls, porous walls, etc. If a system is in an equilibrium state, the latter is largely determined by these boundary conditions.

It is possible to describe the interaction between two systems by assuming that at first they are kept apart by some kind of boundary, which is removed at some later time. In that case the entropy of the combined systems will increase. This has induced Pippard to assume that the entropy is in fact a measure of the "constraints" on the system. If a system is restricted by some kind of boundary condition from reaching a state that it would have if this boundary condition were absent, then its entropy is relatively low.[26] He admits, however, that this view has a severe disadvantage. It overlooks the transient condition between the removal of the constraint and the subsequent arrival at the new equilibrium state. Suppose we remove the constraint, but reinstate it before the system has time to reach equilibrium. Then the system will have an entropy value somewhere between the initial value and the value for the equilibrium state without constraint. Thus the reinstatement of the constraint does not itself decrease the entropy. It just stops the increase of entropy. This implies the impossibility of reducing irreversibility to spatial constraints. As we saw in Chapter 5, the increase of entropy is invariably related to currents.

Nevertheless, Pippard is certainly right in pointing to the relevance of the boundary conditions for irreversible processes.[27] The initial state is also a boundary condition, though not a spatial one. The irreversibility of the physical temporal order makes the initial state relevant, but not the final state. That is, whereas the initial state and the spatial boundary conditions determine physical processes, the final state is merely their effect. Similarly, it is the removal of constraints, not their reinstatement, which leads to a change of entropy.

There is some misunderstanding of the status of boundary and initial conditions. Whereas in kinematics the initial or boundary conditions merely define some otherwise undetermined constants (cf. Sec. 7.3), but are not motions themselves, in physics, the initial and boundary conditions *are* interactions, and are relevant to the subsequent development of the system. There is no depreciation of

26. Pippard 94ff; the Second Law is formulated as: "It is impossible to vary the constraints of an isolated system in such a way as to decrease the entropy" (96).

27. Brillouin Chapter 6.

irreversibility, as a modal law, if we relate it to these physical conditions.

In the context of *statistical physics* the microstate of a system consisting of many molecules is represented by a point in "phase space". A non-equilibrium macrostate is represented by a small domain in this space and an equilibrium macrostate is represented by a large domain. Specifically, a physical interaction creates a non-equilibrium macrostate which is represented by a relatively simply shaped (e.g., spherical) domain in this space.

This is nicely illustrated in a picture in Reichenbach's book.[28] During the spontaneous approach to equilibrium the domain is spread out into a very whimsically shaped "starfish" extending through all phase space, although it has the same volume as the original figure.[29] Any microstate of a system is represented by the set of all positions and momenta of all molecules, which is objectified by a point in phase space. That the initial macrostate must be described by a domain in this space, rather than by a point, is due to the fact that no macroscopic interaction is sufficiently accurate to determine the microstate exactly – no boundary can determine a single point.

The production of a macrostate cannot be understood in kinematical terms only. It needs a new mode of explanation, which is the physical one.[30] To delimit a certain region in phase space requires the introduction of constraints to the positions and momenta of the molecules. As we saw, these constraints have the character, e.g., of a wall – i.e., an external physical influence. In a literal sense, the constraints determine *boundary* conditions, and therefore cannot be so detailed as to provide the exact positions and momenta of all molecules of the system. They cannot determine exactly a point in phase space.

However, if we consider a slight deviation of our macroscopic specification of the state, this will not make much difference to the initial state, nor to the final *macro*state. The point is that the starfish, representing the set of all final microstates, corresponds to the set of all initial microstates compatible with the initial conditions. But this starfish is itself only a very small subset of the set of all microstates representing the final macrostate, because the latter could also have been reached from other initial macrostates incompatible with the actual initial conditions.

28. Reichenbach C 94.
29. This is a consequence of Liouville's theorem. See Tolman 51.
30. Reichenbach C 149ff.

Now suppose we wish to return from the final macrostate to the initial macrostate by reversing all molecular velocities, which is possible in a kinematical sense. This means that with the help of some interaction we have to prepare a microstate falling within one of the "arms" of the starfish. These arms are, however, very thin (because the starfish extends over the whole large domain representing the original final macrostate, but has the same small volume of the domain representing the original initial macrostate). Thus, even a slight inaccuracy in the specification of the reversed final state already means that the process will not end up in the state compatible with the original initial conditions.

This analysis requires that we relate the irreversibility to the accuracy with which a microstate can be prepared by some interaction.[31] Quantum physics has shown that this accuracy has a finite limit determined by Heisenberg's indeterminacy relations. But it is not necessary to appeal to quantum physics. Even in classical physics it is sufficient to state that any physical interaction can only determine *a domain* in phase space. Increasing the accuracy does not make it possible to reduce a domain to a single point. This is a rejection of the classical mechanist doctrine which holds that a physical state can be represented by a point in phase space.[32]

I would like to point out that in this analysis the concept of probability is not used, because it is unnecessary, although the probability of a macrostate is related to its volume in phase space (cf. Sec. 8.5). Sometimes the initial macrostate as described in our discussion is called an "ordered" state, whereas the equilibrium macrostate is assumed to display a maximum disorder (cf. Sec. 6.2). I prefer to delimit the use of the term "order" when related to the law side. Therefore I prefer to speak of a more or less specified state, where "specify" must not be understood in an epistemic sense,[33] but in a physical one, as indicated above.

31. Bondi interprets this to mean that irreversibility is merely due to the inability of the experimenter to produce exact microstates (Bondi, in Gold 3).

32. It may be noted that the probabilistic interpretation of irreversibility also makes use of this fact. The probability of a "point-state" is zero. Only the probability over a domain can have a finite value. Boltzmann's derivation of his famous "H-theorem", which describes the irreversibility of physical processes, also leans heavily on the characterization of the state by a domain (cf. page 166).

33. Cp. Brillouin, Chapter 1, who relates entropy to information. It is true, of course, that in information theory entropy and its increase play an important part, but this has a physical basis.

6.7 *Causality*

In physical texts, the concept of causality is often identified with that of lawfulness, and even with that of determinacy. Therefore, it is discussed with respect to the problem of individuality or the occurrence of stochastic processes.[34] Sometimes, causality is reduced to irreversibility, and conversely, there exist causal theories of time.[35] We shall discuss the law-subject relation and its bearing on determinacy in Chapter 8. Here we want to comment only on the relation between cause and effect.

It is often stated that this relation is ill-defined in physics. It is often impossible to state unequivocally what is cause or what is effect in processes occurring in closed systems. This is not difficult to understand because a study of closed systems requires an initial interest in the interaction as a subject-subject relation. Furthermore, a distinct cause-effect relation is hardly tenable due to the law of action and reaction.

But if we consider external influences on an (otherwise closed) system we can still maintain the causality concept. Especially in this case, we find that the cause-effect relation is irreversible and asymmetric, as it is always understood to be. We speak of causality if the state of a system is changed by some interaction. As such the concept of causality refers back to the kinematical modal aspect. Hence it is an analogical concept, and, as such, it returns in every modal aspect following the physical one.[36]

In order to make this clear, consider the following example of a closed system consisting of two subsystems in thermal contact. Given the respective temperatures, $T_1 > T_2$, a heat current J flows from the first to the second system. This system as a whole cannot be analysed in terms of cause and effect, and physicists will always take recourse to subject-subject relations: the relative energy, the temperature difference, the current. But we can consider parts of the system. For instance, if we consider the first subsystem, we can say that the heat current causes the temperature T_1 to decrease, and if we consider the second subsystem, we can say that the heat current causes T_2 to increase. Alternatively, we can also consider the thermal contact, and state that the temperature difference causes the current J to flow. Thus we find that at the same time the current can be considered both as cause and as effect. This is possible, because in the cause-effect

34. Reichenbach C 55, 149ff; Bunge A, Part I; Campbell A 49-57; Braithwaite Chapter 9.
35. Cf. Whitrow 175, 271f; Frank 53ff; Reichenbach C 24ff.
36. Dooyeweerd B 558, C 110.

relation we neglect the reaction of the system to the cause of its change.

We find, therefore, that the causality concept has a limited applicability – namely, to cases where we can distinguish internal states from external influences. But this does not mean that it is useless, even in physics.[37] Especially in experimental physics, in which external disturbances are deliberately introduced (or at least must be accounted for) in order to study the way a system reacts to them, one frequently makes use of the causality concept.[38]

The main reason why the cause-effect relation is not very useful to physics is that it is not a very simple relation. It is not a subject-subject relation (the effect is not a subject), nor an object-object relation (it is not a succession of states). Whereas the cause is some interaction (a subject-subject relation) in which the reaction is not taken into account, its effect is objective (the changing state of one of these subjects). Thus the cause-effect relation is a complicated subject-object relation, reducible to the basic subject-subject relation called interaction.

37. E.g., Margenau A 389ff; C 437; Toulmin 107ff; see also Bunge A 29, 91ff; Nagel A 25f, B 316ff.
38. Cp. Campbell A 53.

7. Wave Packets

7.1 *Relaxation and oscillation*

In Chapter 5 we discussed the modal retrocipations of interaction: energy, force, and current. These physical concepts referring to the numerical, spatial, and kinematical modes of explanation, are not unrelated. We have seen that currents presuppose forces, and forces presuppose energy. We have also investigated a further complication. We can only give a full account of forces if we consider energy (and other extensive parameters) in opened-up form, i.e., as potentials. And we can only account for currents if we consider forces as fields. Especially in relativity physics, in which the numerical and spatial modal aspects have their kinematical anticipations opened up, the purely retrocipatory concepts of internal energy (or mass) and force can no longer be used, and must be replaced by the energy-momentum fourvector and the field. In this chapter we intend to study the anticipations of the first three modal aspects on the physical aspect more closely.

We start with the opening up of numerical time. We have seen that the latter, originally a purely numerical difference between natural or rational numbers, becomes continuous when anticipating the spatial modal aspect, and uniform when anticipating the kinematical aspect. With respect to physical interaction, it should be subjected to the order of irreversibility.

If for instance two bodies initially at different temperatures are brought into thermal contact, a heat current will decrease their temperature difference. But the heat current in turn is proportional to this difference so that the current will also decrease. The equalization of the temperature will slow down gradually. This process when compared to kinematic relative motion occurs exponentially. It is described numerically by an exponential function, whose exponent is the kinematical time parameter, divided by a constant, the so-called relaxation time.[1] The relaxation time is a measure of the retardation

---

1. This is the time needed to have the temperature difference decrease by a constant factor, namely, the exponential unit ($e = 2.718\ldots$)

between cause and effect. The value of the relaxation time is determined by the conductance of the thermal contact and the heat capacity of the two systems. Therefore it always has a typical and individual character. But the exponential behaviour itself is independent of the typical individuality of the two systems, and is thus of a modal nature.

The relaxation time is not only found in thermal physics. We have relaxation, damping, or absorption in mechanics, whenever there is some kind of friction, resistance, or energy dissipation. We also find it in unstable atomic and nuclear systems: the relaxation or decay of an excited state to the ground state. Relaxation is always related to the transport of energy from one place to another, the transport of energy from one state to another, or the transformation of one kind of energy into another. Relaxation always means the irreversible approach of an equilibrium state.

Oscillation occurs in a system when the equilibrium state is approached with a velocity proportional to the deviation from equilibrium at an earlier instant instead of at the same moment as in relaxation. The system "overshoots" the equilibrium state in an oscillation. One example is a pendulum passing its central (equilibrium) position with a velocity nearly proportional to its amplitude. The amplitude of the oscillation will decrease exponentially due to friction. In fact, oscillation will occur only if the friction is not large enough for a simple relaxation process. The oscillatory motion can be described by a harmonic function, i.e., a sine or cosine function, or an exponential function, which now has an imaginary exponent (see page 44). Besides the relaxation time describing the gradual decrease of the amplitude, we now encounter the oscillation time (the period of the oscillation, the inverse of its frequency) as a typical number. It depends on the internal structure of the system.

Both the oscillation and the relaxation can be used as clocks. In the former case we have to compensate for any kind of relaxation, e.g., for friction in a pendulum clock or a watch. The relaxation time itself can also be used for time measurement, as, e.g., is done in the famous C14 method of determining the age of archeological objects.

Hence we could define a physical time scale, related to the kinematical one by way of the exponential function. The non-linearity of this relation implies that two intervals which are congruent in one of these time scales will be incongruent in the other one. It is, in part, a convention that we prefer to use the kinematical time scale even in physics. But this does not mean that either scale is conventional. In both cases we successfully demand that the kinematical, as well as the physical, temporal relation be properly represented by the scale – namely, the temporal relation must be independent of the typical individuality of the clock by which it is eventually measured (cp. Sec. 3.11).

In a clock based on the physical process of oscillation, the physical aspect of irreversibility is taken care of by the compensation of retardation effects. For a carefully constructed clock the time rate is in accord with the kinematical uniformity of time (the Newtonian metric, see sec. 3.10). The clock must be synchronous to other clocks, which refers to the spatial order of simultaneity. But essentially, the time is measured in a discontinuous way, because the number of periods is *counted*. This shows again that time, as we usually understand this word, is numerical time, opened up anticipating the spatial, kinematical and physical modal aspects.

7.2 *Waves*

The opening up of the spatial modal aspect is realized especially in the introduction of the concept of a field (cf. Sec. 5.5). Fields are intimately related to waves. After Maxwell developed the mathematical theory of electromagnetism, he realized that his equations suggested the possibility of wave motion, which he identified as light. In Chapter 2 we saw that we need to use spatial objects like points and boundaries in order to analyze spatial relations. It took a long time before physicists realized that kinematic objects are needed in a modal description of motion. These kinematic objects are not moving subjectively, but rather provide an objective description of moving subjects.

We have seen in Chapter 2 that real numbers are numerical anticipations to the spatial modal aspect. Real numbers objectify spatial points, real functions of numbers objectify extended boundaries in space, such as lines in a plane or planes in a three-dimensional space. Functions of real or complex vectors also play a role in the anticipations to the physical and kinematical aspects. Especially we shall see that a kinematic subject can be objectified by a set of functions, more or less in the same way as a spatial figure can be objectified by a set of points.

The points on a spatial boundary are connected by an equation. We can say that a function of the form

$$f(x, y): y = ax + b$$

represents the *law* for a straight line in a plane, as long as the numbers a and b are not specified, whereas for certain values of a and b (e.g., $y = 2x + 3$), the equation describes a particular line. From the law we can find a and b if two points on the line (two solutions of the equation) are given. Similarly, the functions in a *wave packet* are determined by a wave equation on the law side, and by specific amplitudes and phase relations on the subject side.

134

Wave packets are kinematical, not physical subjects. They can move, but they cannot as such interact with each other. They are subjects in the first three modal aspects because they are countable and have other numerical characteristics, they have extension and relative position, and they move. But if electrons collide with each other, they do not do so because they are wave packets, but because they are electrically charged and therefore exert a Coulomb or Lorentz force upon each other. This property of being charged is not included in the wave character of the electron's motion – it is an additional property. Two subjects having the same energy but different (eventually no) charge may have similar wave packets. Thus we find that the wave packet is an objective representation of a physical subject with respect to its motion. As such it is of a general, universal, and thus modal character.

Although wave packets are kinematical subjects because they move, the composing waves are not. They are kinematical objects, necessary for the objectification of kinematical subjects. The composing waves do not move and are therefore not subjects. This situation is parallel to the relation between a spatial figure and the points contained in it. The waves composing a wave packet differ from static functions (which anticipate spatial boundaries) by having a time-dependent phase. This phase is responsible for the so-called interference phenomena, which have their static counterpart in the phenomenon of superposition of spatial functions or fields. The superposition and interference properties of waves are extremely important in the description of the interaction of physical subjects. We can say that these concepts anticipate the physical modal aspect.

7.3 *Differential equations*
The mathematical possiblity of describing the motion of a particle with the help of a wave packet was already seen by Hamilton nearly a century before De Broglie stated his famous hypothesis about the wave character of electron motion.[2] It is a direct consequence of the application of differential equations to the problems of motion. The law for a certain motion, whether uniform or accelerated is mathematically objectified in a differential equation[3] whose subjective counterpart is a set of undetermined functions. The equation can only yield a definite solution if some initial or boundary conditions are specified.

This was first recognized by Newton and Leibniz, who inde-

---

2. Jammer D 237ff; Hanson A 450ff; Tolman 42.
3. Margenau A 182.

pendently invented differential and integral calculus in order to be able to study mathematically the motion of material bodies.[4] Thus the mathematical expression of the law of purely kinematical motion, Newton's first law, is (in Leibniz' notation)

$$\frac{d\mathbf{r}}{dt} = \mathbf{v}$$

The solution of this equation, $\mathbf{r}(t) = \mathbf{r}(0) + \mathbf{v}t$, contains the undetermined parameters, the initial position $\mathbf{r}(0)$ and the velocity $\mathbf{v}$. If they are known, the position of the moving body is given for any time. The moving subject itself is assumed not to change and its position is therefore represented by a characteristic point, e.g., its centre of mass. This law is valid for non-interacting subjects provided $\mathbf{r}$ and $\mathbf{v}$ have reference to an inertial system.

By differentiating the equation, we find

$$\frac{d^2\mathbf{r}}{dt^2} = 0$$

as an equivalent expression of the same law. If we now introduce interaction in the form of a force or field $\mathbf{F}(\mathbf{r})$, the second law of motion is

$$\frac{d^2\mathbf{r}}{dt^2} = \mathbf{F}(\mathbf{r}).$$

The solutions of these equations do not describe motion itself, but the spatial path of the motion – i.e., a retrocipatory spatial analogy of kinematical motion.[5] But what we need is an anticipatory description. We have already found that a function $f(\mathbf{r})$ is a numerical anticipation of spatial figures. Therefore we should subject functions of this kind to a differential equation, rather than point vectors $\mathbf{r} = (x, y, z)$. We must then differentiate with respect to all temporal and spatial co-ordinates (t and $\mathbf{r}$). This can be done in several ways.

For electromagnetic wave propagation in vacuum, Maxwell found the following law as a consequence of his laws concerning electric and magnetic fields:

$$\nabla^2 f(\mathbf{r}, t) = \frac{1}{c^2} \frac{\delta^2 f(\mathbf{r}, t)}{\delta t^2}$$

where ∇ is the differential operator $\left(\frac{\delta}{\delta x}, \frac{\delta}{\delta y}, \frac{\delta}{\delta z}\right)$ and $f(\mathbf{r}, t)$ represents the electromagnetic field.

4. Beth A 132ff.
5. The problem of finding the path of the motion is elaborated in the Lagrangian representation of mechanics.

The solution of this equation depends on the boundary conditions in a rather complicated way. If there are no boundary conditions specified, any function of the type $f(r \pm ct + \varphi)$ is a solution of this equation.[6] This shows that the number c is the velocity of the kinematic subject whose motion is described by the equation. The velocity c belongs to the law of the motion and is not dependent on initial or boundary conditions. This wave equation is relativistically invariant, and c has the same value with respect to any inertial reference system. Consequently, the wave equation can only describe the motion of subjects whose internal energy (or rest mass) is zero: light quanta or neutrinos. Another difference with Newton's law is that $f(\mathbf{r}, t)$ does not describe the path of the motion.

Another example is Schrödinger's equation:

$$\nabla^2 f(\mathbf{r}, t) = 2im \frac{\delta f(\mathbf{r}, t)}{\delta t}$$

where i is the imaginary unit, and m is a constant, later to be identified with the mass of the subject (if it is physically qualified). The solution is of the type $f(i\omega t \pm i\mathbf{k}.\mathbf{r} \pm i\varphi)$. The main difference with Maxwell's equation is that it applies to a physical subject having mass and moving with a low velocity compared to c. It is not relativistically invariant, whereas its solutions are complex functions. ω and $\mathbf{k}$ are determined by the boundary conditions. Similarly as with Newton's equation, we can add other terms describing motion in an external field.

These are not all the possibilities for differential equations describing motion. For instance, we can write the Schrödinger equation in a relativistically invariant form, which gives us the Dirac equation or Klein-Gordon equation. Currents are also subjected to differential equations. In classical physics one distinguished particle motion from a continuous current. The wave theory of motion shows that this distinction is unwarranted. Particle motion also achieves the character of a current when anticipating the physical modal aspect.

On the other hand, any physical current must be quantified when taken in the retrocipatory direction, i.e., when its energy is involved. In so far as the Maxwell and Schrödinger equations do not show damping, they are idealized limiting cases of real physical motion.

---

6. φ is an arbitrary unspecified number. It is sometimes called the *phase*, but just as often this name is used for the whole argument $r \pm ct + \varphi$. Note that $r = |\mathbf{r}|$.

7.4 *Superposition*

Maxwell's and Schrödinger's equations also differ from Newton's equation because they are homogeneous and linear. This means that if we have two different solutions, f_1 and f_2, any linear combination of them $(af_1 + bf_2)$ is also a solution. Here a and b are arbitrary real numbers for Maxwell's equation, and complex numbers for Schrödinger's equation. This implies that we can order the solutions in a "function space", if we can design an operation called the scalar product, with properties as mentioned in Sec. 2.4. The basis of this function space depends on the boundary conditions. Presently, we are interested in the unbounded case of uniform linear motion. The basis functions in this case form a continuous set. This presents a difficulty because these functions cannot directly be normalized, but this problem can be solved, as we shall see below.

For the Maxwell equation this basis consists of the functions

$$\cos(\omega t \pm \mathbf{k}.\mathbf{r} + \varphi)$$

where ω (the angular frequency) and $\mathbf{k}$ (the wave-vector) range over all positive and negative real numbers, with the condition

$$\omega/|\mathbf{k}| = c$$

For the Schrödinger equation the basis is formed by the set of complex exponential functions

$$\exp i(\omega t \pm \mathbf{k}.\mathbf{r} + \varphi) = \cos(\omega t \pm \mathbf{k}.\mathbf{r} + \varphi) + i \sin(\omega t \pm \mathbf{k}.\mathbf{r} + \varphi)$$

with the same range for ω and $\mathbf{k}$, but without the restriction of Maxwell's equation. It is clear that the two cases are very similar. In both cases, φ depends on ω and $\mathbf{k}$, but not in a regular way. Just as the amplitude $A(\omega, \mathbf{k})$, it is determined by the initial conditions. Because

$$\cos(\omega t \pm \mathbf{k}.\mathbf{r} + \varphi) = Re \exp i(\omega t \pm \mathbf{k}.\mathbf{r} + \varphi)$$

where "Re" denotes "the real part of", we can, from now on, consider the solutions of the two equations simultaneously, if, in the case of Maxwell's equation, we add the prefix "Re" and consider the amplitude as a real number.

Because the basis functions form a continuous set, they must not be summed but integrated. Hence any solution of the wave equation can be decomposed into the basis wave functions:

$$f(\mathbf{r}, t) = \int A(\omega, \mathbf{k}) \exp i(\omega t \pm \mathbf{k}.\mathbf{r} + \varphi) \, d\omega d\mathbf{k}$$

The scalar product is now defined by

$$(f_1, f_2) = \iint A_1{}^*(\omega, \mathbf{k}) A_2(\omega', \mathbf{k}') \exp-(\omega, \mathbf{k}) \exp+(\omega', \mathbf{k}'). \\ .d\omega d\omega' d\mathbf{k} d\mathbf{k}'$$

which shows that the norm of the function f is given by

$$\| f \|^2 = (f, f) = \int A^*(\omega, \mathbf{k}) A(\omega, \mathbf{k}) \, d\omega d\mathbf{k}$$

This norm can now be normalized to give the value 1 for an elementary physical subject.

Thus we find that the function describing the motion of a free particle can be decomposed into a set of "plane waves", each of which is characterized by a frequency ω and a wave vector $\mathbf{k}$. All functions have these plane waves in common, but different values for $A(\omega, \mathbf{k})$ and φ correspond to different functions.

As long as we stick to kinematics, we can say nothing more about the values of the amplitude and the phase. For freely moving, physically qualified systems these values are determined by their latest interaction – by the way they were "prepared" before they started their free motion. This preparation can be understood as the result of a collision with another system, by the "birth" of the system while emerging from another one (as in the case of a light quantum emitted by an atom), or in an instrumental sense: light or electrons passing a shutter. In all these cases, the particle's spatial extension will be determined by that interaction, as well as its "temporal extension" which is the duration of the interaction – e.g., the time the shutter was open, the relaxation time of the emitting atom, or some other characteristic time. Immediately after the particle has started its free motion, the "temporal extension" can be understood as the time needed to pass a certain point. It is related to the spatial extension by means of the velocity of the particle with respect to that reference point. For particles having subluminal velocities, the spatial extension of the wave packet will always increase.

This means that a single plane wave, although it is a solution of the wave equation, cannot serve as a representation of a kinematical subject. The plane wave cannot even be said to move. Because it is infinitely extended, its appearance varies periodically, but there is no displacement. It also cannot be normalized. This is not serious, just because it will not be used to represent a moving subject, and because a wave *packet* consisting of all plane waves can be normalized by a proper choice of the amplitudes and phases. This is important for the interpretation in which wave packets describe probabilities with respect to future interactions (see Chapter 8). We should, therefore, consider plane waves as *objects* in the wave packet.

The wave packet formalism is not only relevant to the motion of a free particle, or of a particle in a field (in which case the Schrödinger equation must be adapted), but also to currents. In fact, Fourier developed the above theory (named after him as Fourier analysis) while studying the problem of heat conduction. Also light quanta are individualized currents in the electromagnetic field.

The velocity of the wave packet is its "group velocity", $d\omega/d|\mathbf{k}|$. Usually, ω and $\mathbf{k}$ are not independently variable. For wave packets subjected to Maxwell's equation, $\omega = c|\mathbf{k}|$, hence the group velocity is c, independent of the reference system. If ω is not proportional to $|\mathbf{k}|$, it is said that the waves show "dispersion". For uniformly moving material particles, satisfying Schrödinger's equation, ω is proportional to $|\mathbf{k}|^2$. Hence, the group velocity (the particle velocity) is proportional to $|\mathbf{k}|$, and depends on the choice of the reference system.

7.5 *Energy and momentum*

Now we have to emphasize the fact, already mentioned in Sec. 2.4, that the set of plane waves which serves as a basis for the solutions of the wave equation is by no means unique. There is an infinitude of alternative bases, one of which we shall discuss in Sec. 7.7. On the one hand it is a bit unfortunate that we nearly always take the set of plane waves as a basis, because it suggests that this is somehow intrinsic to kinematical and physical subjects, which is not the case. On the other hand, there are, of course, good reasons to single out this basis. One reason is that the exponential functions are rather convenient in a mathematical sense.

However, there is a more fundamental argument. We have started our discussion with the problem: how can we describe uniform kinematic motion? This problem presupposes the isotropy and homogeneity of space and time. Hence we could restate the problem in the following way: find a complete set of basis functions which reflect the temporal and spatial isotropy and homogeneity. Or, in group-theoretical terms: find a suitable representation of the Galileo group (or, eventually, of the Lorentz group). One then finds the set of plane waves, which is therefore the most natural basis set for the description of uniform motion. This immediately implies that if space, e.g., is not homogeneous (e.g., in the presence of a central field of force) the plane waves, though still possible, do no longer form the most suitable representation of a physically qualified subject.

Now we have seen (Sec. 5.3) that the energy-momentum four-vector of a freely moving physically qualified subject must be a (frame dependent) constant because of this assumed temporal and spatial homogeneity. In a similar way we also find that the characteristic frequency and wave vector of a wave packet form a (frame dependent) constant for exactly the same reason. However, a theorem developed by Noether states that each type of symmetry has one and only one such constant. Therefore, the energy E must be proportional to the frequency f or $\omega = 2\pi f$, and the momentum $\mathbf{p}$ must be proportional to the wave vector $\mathbf{k}$, if these objective properties all belong to the same physical system. The proportionality constant is determined only by the choice of units for these variables, and is therefore a general, modal, universal "constant of nature". It is known as

Planck's constant, h or $\hbar = h/2\pi$. The fact that it has the same value for all subjects anticipates the possibility of all physical subjects interacting with each other.[7] Hence, $E = \hbar\omega = hf$, and $\mathbf{p} = \hbar\mathbf{k}$.[8]

These relations (of Planck and De Broglie, respectively) imply that the frequency and the wave vector are connected by

$$hf = \hbar\omega = \hbar^2 k^2/2m \quad \text{because} \quad E = p^2/2m$$

where m is the, up till now, unspecified constant in the Schrödinger equation (which was designed such as to give this result).[9] The constant m can now be identified with the mass of the subject, whereas its velocity is equal to $\hbar\mathbf{k}/m$.

The nature of Planck's and De Broglie's relations does not imply that energy/momentum and frequency/wave vector are conceptually identical. The former is retrocipatory while the latter is anticipatory. But these two directions in the intermodal relationships are always strongly related (remember that there is a one-to-one correspondence between the real numbers and the line segments on a straight line). The proportionality of energy and frequency means that energy is not related to the amplitude of the waves, as was assumed earlier, so that we have to find another interpretation for the amplitude (cf. Sec. 8.6).

The proportionality of energy and frequency is sometimes misunderstood as "energy quantization". In a light beam of frequency f we can only have particles of energy hf, which seems to imply discreteness. But the "discreteness" is due to the starting point (a light beam of frequency f). Energy is a variable which has a continuous spectrum, as does frequency, for freely moving subjects. The energy variable in classical physics also has a continuous spectrum. Only in bounded systems, such as atoms or molecules, can the internal energy spectrum become discrete.

There are some speculations in the literature about a possible quantization of physical space and time.[10] But this supposed quantization must not be misunderstood. It simply means that there is perhaps a smallest distance (called hodon) and a smallest time interval (called chronon) by which one can distinguish subjects or events through physical means. It does not mean that any distance or time interval is just an integral number of this hodon or chronon, re-

7. Messiah 149.

8. The isotropy of space gives rise to the intrinsic spin of any particle, cf. Sec. 9.7.

9. The Schrödinger equation as given on page 137 must be slightly adapted to account for the occurrence of h.

10. Margenau A 150ff; Jammer A 184; Russell B 42.

spectively. This would certainly lead us into antinomies. For instance, the diagonal of a square would be equal to its sides.

7.6 The Heisenberg relations

The shape of the wave packet as determined by its preparation is mathematically described by the set of amplitudes $A(\omega, \mathbf{k})$ and phases $\varphi(\omega, \mathbf{k})$, such that the net amplitude is only appreciable within the packet, whereas the composing waves add up to zero outside it. If we denote the spatial extensions by $\Delta x, \Delta y, \Delta z$, and the temporal extension as defined on page 139 by Δt, we find that we need waves with non-zero amplitudes for frequencies and wave vectors within a certain range of values around a central value. If we denote these ranges by $\Delta \omega$ or Δf, and by $\Delta k_x, \Delta k_y, \Delta k_z$, we find by very general reasoning that [11]

$$\Delta\omega.\Delta t \geqslant 2\pi \text{ or } \Delta f.\Delta t \geqslant 1, \text{ and } \Delta k_x.\Delta x \geqslant 2\pi, \text{ etc.}$$

This means that although a wave packet can be characterized by a certain frequency f and wave vector $\mathbf{k}$, this is not a precise characterization as in the case of a single plane wave. The spread in the values of $\mathbf{r}$ and t determines the spread of $\mathbf{k}$ and f, in the sense, that if $\mathbf{r}$ and t are precisely determined because the packet is small, $\mathbf{k}$ and f are ill determined, and conversely. These relations for wave-based signals (as occur in electric communication systems) were already derived by Heaviside, long before Heisenberg introduced them in quantum physics. They are of a general, modal character, not characteristic of the typical structure of any physical system – in fact, they have a kinematical meaning.

Because of the relations of Planck ($E = hf$) and De Broglie ($\mathbf{p} = h\mathbf{k}$), we have

$$\Delta E.\Delta t \geqslant h \text{ and } \Delta p_x.\Delta x \geqslant h, \text{ etc.}$$

These are the so-called Heisenberg relations. They say that the energy and the momentum of a particle are not exactly determined, because of the wave character of their motion.

The first mentioned (so-called fourth) Heisenberg relation is sometimes criticized because (in contrast to the other ones) it cannot be derived from a relation between non-commuting operators (cf. Chapter 9).[12] However, operators play no essential role in the kinematical theory of wave motion and we do not need operator calculus in order to derive the above result.[13]

It will not immediately be clear that the wave description with the inherent Heisenberg relations, is also valid with respect to systems

11. Heisenberg A; Jammer D 323ff.
12. Bunge B 267ff.
13. Messiah, Chapters 4 and 8.

with high energies – e.g., fast particles in a bubble chamber, or macroscopic bodies. One must first realize that the spread in the Heisenberg relations is not measured relative to the energy or momentum of the subject itself. Hence the spread in energy of a high energy system can be very large compared to the spread of a mono-energetic electron, and still be extremely small with respect to the total or kinetic energy of the system itself. The former means that the system can be sharply localizable (both temporally and spatially), while the latter means that the subject apparently has a very precise value for its energy, because the spread is so small relative to the total energy. The wave phenomena become determinable only if the spread as determined by Heisenberg's relations is comparable to the energy, relative position, etc., of the subject itself. Thus, in principle, a planet's motion must also be described as that of a wave packet, but, as yet, there are no experiments to show this. Nevertheless, the wave theory is not limited to small subjects, and is therefore of a general, modal character.

On the other hand, it has special consequences as soon as the momentum p, for example, is of the order of Δp. If an electron is restricted to a limited spatial region (e.g., a hydrogen atom) the mean value of p has a smallest value determined by Heisenberg's relations, and thus a smallest energy. If this spatial region can be extended (e.g., if the electron no longer belongs to one hydrogen atom, but to a molecule of two atoms), the electron can decrease its momentum, and thus its energy. This "exchange bonding" or "covalent bonding" explains why hydrogen is a diatomic molecule.[14] By a similar argument we can explain why an electron cannot exist as an independent particle in an atomic nucleus. Its total energy as determined by Heisenberg's relations would be more than its rest mass. The much heavier mesons can exist independently in a nucleus for a short period of time.

Another consequence of the Heisenberg relations is already mentioned in Sec. 5.3. If a system is isolated only during a short time Δt, the conservation law of energy has a restricted validity. Energy is now constant within the limits of $\pm \Delta E = \pm h/\Delta t$. For macroscopic systems, this amount is unmeasurable small compared to the total energy E. But this inaccuracy has detectable consequences for some subnuclear processes. Also if the "life time" of some excited state is Δt, the energy of this state is only determined within the accuracy $\Delta E = h/\Delta t$.

14. The formation of molecules with more than two atoms is energetically less favourable because of Pauli's exclusion law.

7.7 *Interference and Huygens' principle*

A wave packet is a superposition of waves with different frequencies and wave lengths (the wave length is inversely proportional to the absolute value of the wave vector). We speak of interference of waves if we add waves of the same frequency. In that case we can simply add the amplitudes of the waves in the manner of complex numbers, i.e., by taking into account the phase relations. The phenomenon of interference is the basis of Huygens' principle, according to which a propagating wave signal can be decomposed into spherical waves. Every point of space is assumed to be the centre of an expanding wave. The actual motion of the signal is the superposition of all these spherical waves with their different amplitudes and phases, which are thus determined by the initial and boundary conditions, as is the case with the above mentioned plane waves. This illustrates the arbitrariness of the choice of the basis for the decomposition of an actual wave packet.

The spherical waves used in this case are less easy to handle mathematically. E.g., it is difficult to prove that light moves approximately rectilinearly and in one direction. This was not done until the 19th century by Fresnel and Kirchhoff. This makes it understandable why Newton's corpuscle theory of light propagation was favoured over Huygens' theory for over hundred years. Finally, the experiments of Young and Fresnel proved the possibility of interference which cannot satisfactorily be explained in Newton's theory. In addition, Fizeau showed that Newton's theory gave a wrong value for the speed of light in a medium.

We have seen that the plane wave representation is favoured in the description of pure kinematic motion because it reflects the temporal and spatial homogeneity and isotropy which is assumed. In Huygens' theory only temporal homogeneity is assumed. Therefore, the frequency is still related to energy and remains invariant. But, because every spherical wave has a singular point, it lacks spatial homogeneity, and therefore there is no relation to linear momentum. Thus Huygens' representation is especially fruitful for the description of the wave's interaction with rigid bodies in a spatial sense: reflection against a wall, refraction through a boundary between two media in which the velocity is different, diffraction by a slit in a wall (or a hole, or several slits, or a grid). In all these cases the physical details of the interaction are neglected. Only the change of motion is considered, by the study of the effect of the interference of the spherical waves in the neighbourhood of these spatial structures.

Huygens' principle is very successful in the solution of problems of this kind, most of which cannot be solved on the basis of Newton's mechanics. In fact, it is mainly because of these phenomena

144

(especially diffraction) that the wave theory of motion is accepted. The physical community has become especially convinced of the correctness of the wave theory through interference phenomena. Interference causes photons or electrons to be in positions unexpected by Newtonian mechanics (and conversely, these entities are not present at positions expected by Newtonian mechanics).

Plane waves must first be decomposed into spherical waves before we can analyze such experiments. This is possible (and also the reverse: the decomposition of spherical waves into plane waves) because the spherical waves, as well as the plane waves, form complete sets, such that they can serve as a basis for the decomposition of the solutions of the wave equation. These two possibilities are not exclusive and are only two instances of an infinitude of possibilities. Thus in atomic and solid state physics one often uses a more limited set of plane waves, spherical waves, or even combinations of them.

More than anything else, Huygens' principle shows the anticipatory character of the wave theory of motion. It can only manifest itself in the interaction of the particle with a rigid body, but in the mathematical description (which is only concerned with the motion of the particle) one completely abstracts from all the physical details of this interaction. The wave theory is a kinematic theory, opened up towards the physical modal aspect.

7.8 *The wave-particle duality*
The distinction of waves as objects and wave packets as subjects in the kinematic modal aspect shows that there is not really a wave-particle duality in a kinematical sense. Any physical and kinematic subject can only be represented as a wave packet, which has some of the characteristics of a particle – namely, a more or less precise position, momentum and energy. However, since the classical doctrine maintains that all physical phenomena must be reduced to the motion of unchangeable pieces of matter, an elementary particle is defined (or rather deified) as having definite values at a particular time for its position, energy and momentum. This philosophically coloured idea of a particle clashes with the concept of a wave packet. This is the reason why wave theory and especially the Heisenberg relations have been the subject of so many discussions.[15] We have to stress that the wave packet is only a modal and anticipatory description of moving, physically qualified subjects. Therefore it is limited in two respects.

First, the wave packet does not describe interactions in a subjective

---

15. See, e.g., Bohr A; Jammer E, Ch. 3; Klein B; Margenau A, Ch. 16; Reichenbach D, Ch. 11; Price, Chissick.

sense. This is in sharp contrast to the classical concept of moving particles, whose extension was supposed to be impenetrable (wave particles are far from that). The assumed impenetrability gives rise to the possibility of collisions between the particles – in fact, the only kind of interaction admitted in atomistic classical mechanics. Huygens' principle only enables us to give an objective description of the kinematical consequences of a very simple kind of interaction – namely, an interaction in which the wave packet collides with a rigid spatial system, and all physical details are disregarded. Hence, the collision between two atoms can be described by the wave theory only after the typical structure of the interaction is "translated" into spatial terms (the collision cross section). However, real interactions, especially those in which the internal state of one of the systems is changed (e.g., if a system is absorbed), cannot be understood within the framework of wave theory alone.

Secondly, the wave theory gives a *modal* description and therefore discards all *typical* properties of the described subjects. It is completely irrelevant whether we are dealing with electrons or light quanta if we apply Huygens' principle. The diffraction patterns made by a beam of light or by electrons of comparable wave lengths passing through a hole or a crystal are perfectly similar. This was predicted by De Broglie in 1923 and confirmed soon afterwards by Davisson and Germer.[16]

The kinematic character of diffraction and interference also manifests itself in the two-slit experiment in which a wave packet is split up. Interference of the two parts occurs as soon as they meet each other again.[17] It should be emphasized that in a kinematic sense the splitting of a wave packet into two parts (after passing a screen with slits, for instance) is no problem. It appears that after this transition we have two wave packets, two subjects spatially divided. Indeed, in a spatial sense, we cannot speak of one subject, if its parts are not connected. But in the case of wave packets, we have kinematic subjects, and its parts must not be spatially, but kinematically connected. Indeed, the two parts of the wave packet, after passing the double slit, are kinematically coherent. The well-known interference phenomena are explained by assuming that the two parts of the wave packet have well-determined phase relations.

The diffraction experiments especially emphasize the fact that the wave theory cannot account for the individuality of the particles and their individual interactions. We must remember that the waves,

16. Jammer D 246, 251; Klein A.
17. See on interference experiments Feyerabend C 199ff; Jauch 112ff; Bohr A; Fine; Reichenbach B 24-32.

rather than the particles, interfere in diffraction, reflection or refraction. This was not always clearly recognized. At first, some people tried to explain these phenomena by assuming that different particles interfere. But experiments soon showed that, if one has a very dilute beam with only one particle at a time in the apparatus, one still has the same diffraction patterns.

Thus one assumed that the waves in a single particle interfere with each other. But even this is objectionable. For example, interference between the beams emerging from two lasers is possible. In this case one can also dilute the beams such that no particles are present in the apparatus roughly 90% of the time and one particle is present roughly 10% of the time. The interference phenomena were decisively different if both lasers were open or if one was closed. Thus a particle emerging out of one laser interferes with the field of the other one, even when there are no particles coming out of the latter.[18]

The wave theory itself cannot give a full account of the individual behaviour of the described subjects. We have to supply this theory with an interpretation which is no longer of a purely modal character. On the one hand, we have to give a probability interpretation of the waves describing the motion of the particle. As we shall see, the theory of probability is also an anticipatory one, which explains its strong connection with wave theory (cf. Sec. 8.6). On the other hand, we must show that the physical concept of a particle refers to a typical structure (cf. Chapter 10).

18. Cf. Pfleegor, Mandel.

8. Individuality and Probability

8.1 *Individuality*

We have now studied the first four modal aspects and their retrocipations and anticipations both on the law side and the subject side. Thereby we have outlined an answer to what we consider the first basic problem of science (Sec. 1.2): *Are there general modes of experience which provide an order for everything within the creation, and if so, which are these universal orders of relation?* If we were only interested in mathematics we could finish here. But the physical modal aspect is one aspect – the first one – by which certain typical structures are qualified, and in which one has to account for individual phenomena. This means that physics is not only concerned with modal laws and subjects, but must also study special laws like those of electromagnetism, and typical structures like that of the copper atom.

In other words, physics is also confronted with our second basic problem: *How can stable things exist, and how can they change?* Before we enter into this study, we have to pay attention to the law-subject relation for individual systems. This leads us to the theory of probability. Statistics applies the theory of probability to the properties of a collection or ensemble of systems *with the same typical structure*.[1]

The necessity of using probability theory in physics was not always recognized. Until the beginning of the 20th century, classical mechanics served as a deterministic prototype of the physical sciences. We already noted that mechanics is a modal theory. It is almost exclusively concerned with motion as a mode of being of physically qualified subjects. Abstraction took place on the subject side from all concrete properties which do not relate to the kinematic aspect. Each concrete thing is thereby reduced to a modal kinematic subject. Because it remains physically qualified, nevertheless, the retrocipatory aspects of interaction (mass, energy, force) have to be included. The simplest objects of mechanics are "mass points", with forces acting between them.[2]

1. Tolman 2, 43.
2. Einstein B 19ff.

In a deterministic interpretation this kinematic aspect is absolutized. All other aspects, which together with the kinematic one determine concrete reality, are ignored or dismissed as "secondary qualities". This is especially the case with individuality, which though not lost completely, has a mere rudimentary existence as motion-subjectivity. When mass, position, velocity, and external circumstances (seen as forces or force fields) are given in a specific point at a certain time, motion is fixed with relation to past and future. Even contemporary authors characterize "particles" as being "localizable".[3]

On the law side a correction of this rigorous functionalistic determinism was offered by classical chemistry, whose basis was laid by Priestley, Lavoisier, Dalton, Berzelius and others at the end of the 18th and the beginning of the 19th century. It differed from mechanics in that it ascribed typical properties to its objects, the elements consisting of similar atoms, and the chemical compounds consisting of similar molecules.

In physics a merely modal, deterministic approach first began to fail on the subject side. The individuality of atoms and molecules made its entry, first in statistical mechanics, then in radioactivity, and finally in the Brownian motion. In chemistry essentially probabilistic reasoning underlies the law of mass action established by Guldberg and Waage.

However, both chemists and physicists still believed in determinism. Statistical methods were only used for practical reasons because a fully deterministic calculation of the motion of the many particles constituting a gas was (and is) beyond human capabilities.[4] Although radioactivity was considered to be a mystery, at the turn of the century physical scientists were still confident that it could be solved along deterministic lines, i.e., by a modal theory.

All this changed as a result of the development of quantum physics in which better distinction is made between an individual system and its state. This state has, in a certain sense, a latent character for an isolated system, manifesting itself only if the system interacts with another one – e.g., with a measuring apparatus. According to quantum physics, the individual state of the system does not exactly determine the result of the interaction. The initial and final states of the system before and after the interaction are not related in a purely modal, determined way, but by means of a probability law.

This so-called stochastic relation is therefore not lawless. The probabilities of the joint initial and final states as numerical pre-

3. Bunge B 108, 24; Akhieser, Berestetsky 17; Messiah 4, 138; Čapek, Ch. 14.
4. Reichenbach C 56.

dicates of possible interactions are determined by the typical structure (the law) for the interacting systems. There are many different interpretations of this state of affairs,[5] three of which we shall briefly discuss.

A small number of physicists (among others, Einstein,[6] Schrödinger, Bohm,[7] and the school of De Broglie) remain loyal to determinism and therefore hypothesize the existence of (as yet) unknown determining factors (called hidden variables). In his mathematical analysis of quantum physics, Von Neumann[8] has shown that hidden variables cannot weaken the indeterministic structure of quantum physics (if the latter is correct). Physicists who still consider determinism, or rather a purely modal theory, as exclusively acceptable, are forced to assume that although the quantum physical formalism accurately describes the phenomena, it is nevertheless incorrect or incomplete. In principle this view cannot be contradicted, but it is not very convincing as long as its proponents have not succeeded in designing a theory along these lines.[9]

The large majority of physicists emphasize the measuring process.[10] According to this view, we do not really know anything about a closed system. Only the results of measurement are verifiable, and during measurement the examined system cannot be isolated. But the result of measurement is not only determined by the character and the state of the system, but also by the action of the measuring instrument: this is called the measurement disturbance.[11] Taken by itself, this phenomenon is not invented in quantum physics, of course. In classical physics one also knew about these errors in measurement, but physicists believed that in principle the measurement disturbance could be made arbitrarily small. In quantum physics, this is no longer tenable. The discovery that all moving subjects must be described with the help of wave packets implies that measurement disturbance cannot be arbitrarily small.

According to the so-called Copenhagen Interpretation[12] – of which there are several variants – it is quite possible that an isolated system is completely determined. However this is considered to be a meaning-

5. The literature on the interpretation of quantum mechanics is very extensive. For a review of the literature, see Dewitt, Graham; Scheibe; Hooker.

6. Einstein B 82ff; Klein A, B; Hooker.

7. Bohm.

8. Von Neumann; see also Jauch, Chapter 7; Jammer D 366ff, E 265ff.

9. On the completeness of quantum theory, see Jammer D 366ff.

10. Cp. Bohr B, Introduction: "The aim of science is to extend as well as to order our observations . . .".

11. See, e.g., Tolman 16, 17.

12. Heisenberg C, Chapter 3, 8; Losee B; Hanson A.

less proposition because it is not experimentally verifiable.[13] Within this concept, the problem of the individuality of physical systems is disposed of as an epistemological problem about the relation between observing subject and observed object.

In Bohr's writings[14] one finds some remarks to the effect that "... a not-further-analyzable individuality ... has to be attributed to every atomic process ..."[15] According to Meyer-Abich,[16] the individuality of atomic processes belongs to the heart of Bohr's interpretation of quantum physics. But, as he makes clear, in Bohr's view this individuality arises from the (human) subject – (sub-human) object relation. Bridgman also assigns a large role to the act of observation in the interpretation of quantum physics.[17]

It is true that we wish to observe the object of our investigation in a measurement. Observation is a psychically qualified human act, which also has a physical aspect. But it is only this physical aspect which we need to take into account in our discussion of the limitations of measurement. It arises from the interaction between the object of measurement and the measuring instrument (eventually our own senses). As Bridgman rightly observes,[18] this implies that the object cannot be considered isolated. Theoretically, we prefer to study isolated systems, but in measurements we observe a system while it is interacting with the measuring instrument. In this interaction, however, we do not have to consider a subject-object relation of observer and observed system, but a subject-subject relation of two interacting physically qualified systems.[19]

According to a third interpretation the state function is not related to a single system, but represents the way in which a large number (an "ensemble") of similar systems is prepared.[20] It is possible to determine the state function by means of a large number of measurements on the ensemble, but this procedure is meaningless for a single system, whose individuality must be ignored. The state function is an expression of our knowledge of the ensemble.

All three interpretations emphasize undeniable states of affairs. Two aspects which they all have in common can be criticized. In the

13. Cp. Heisenberg B 29: "... *dass die unvollständige Kenntnis eines Systems ein wesentlicher Bestandteil jeder Formulierung der Quantentheorie sein muss.*" For a criticism of this view, see Popper C.
14. Bohr A 209, 223, 230.
15. Jammer D 347.
16. Meyer-Abich 102.
17. Bridgman, in Henkin 227ff.
18. Bridgman, in Henkin 229.
19. Cp. Čapek 303f.
20. See, e.g., Groenewold, in Bastin 43-54.

first place, they all refer, implicitly or explicitly, to the deterministic interpretation of classical physics without being sufficiently aware of its philosophical bias. In the first interpretation, the determining factors are taken to be unknown as yet. In the second it is posited that they cannot be measured if they exist. In the third one takes recourse to the "ensemble" because it is assumed to be fully determined. Hence there is, in effect, no break in principle with 19th century determinism. For instance, Heisenberg posits that only its premiss is invalid, i.e., the premiss: "If at a certain moment position and velocity of all particles are known." Heisenberg therefore does not consider determinism as incorrect, but rather unapplicable in quantum physics.[21]

Secondly, the mathematical formalism of physics, in fact, does not receive its due. It is generally accepted that the theory has a statistical character. The first interpretation mentioned above does recognize that the formalism describes the phenomena accurately, but it refuses to accept the conclusion that physical phenomena themselves have a stochastic character displaying individuality. The second interpretation misses the point that measurement disturbance has no significance for the calculation of the probable measurement results. The third interpretation ignores the fact that the mathematical formalism ascribes a state function to each separate system. Moreover, it must be observed that the application of statistical laws, for example, in quantum physics with respect to radioactivity assumes that the decay of different atoms constitutes statistically independent events.[22]

The underestimation of the mathematical formalism is not so strange because the formalism is generally considered to be merely a handy framework within which empirically discovered physical law structures can be summarized. After all, is not mathematics a free creation of the human mind? This is true as far as mathematics is a theoretical opening up of some modal aspects of temporal reality. But these are modal aspects of concrete reality which make its understanding possible. The mathematical formalism of quantum physics is more than a convenient representation of our knowledge of inorganic structures. It is the theory regarding their mathematical aspects, and an objectification of their physical aspect.

Quantum physics does not prove that individuality may be attributed to physically qualified subjects, but it leaves room for such a conclusion. No special science can solve this philosophical problem. A

21. Jammer D 330, E 75ff; see also Heitler 192.
22. Cp. Hempel B 392.

scientific theory, seeking as a matter of course to stay close to empirical concrete reality, is able to display a deterministic structure which excludes the possibility of individuality, but is also able to leave room for individuality. The former is the case with classical physics while the latter exists in modern physics.

In itself it is correct that science takes distance from the individual. Inherently, science involves abstraction, and the first abstraction to be made is one from individuality. A solid state physicist will do many experiments with a single crystal, yet his interest is not directed to this one crystal, but extends either to the modal physical laws to which the crystal is subjected or to its typical structure. In the analysis of the results of his measurements he constantly abstracts from the subjective individuality of the object of measurement. In this respect quantum physics disregards individuality as much as classical physics did.

However, it has become necessary to account for the fact that natural phenomena cannot be completely described in a deterministic way. This is a philosophical matter, and before one can start its analysis, one has to make a choice concerning the individuality of natural subjects, whether it will be accepted as a matter of fact or not. Thus, Van Melsen says: "The assumption of determinism in matter is less result of than condition for science."[23] He rejects the subjective individuality of e.g. radioactive particles, ". . . each having separate existence . . ."[24] After posing the dilemma: Natural necessity (fully determined by law) or chance (in the sense of absolute arbitrariness), he rejects the latter.[25]

But we reject the dilemma,[26] replacing it by the *correlation* of law and subject, which cannot be reduced one to the other. Determinism reduces the subject to the law while pure chance eliminates the law. In our view, individuality is not an afterthought, a result of a conclusive analysis, but a premiss for understanding physics. This also implies the distinction of typicality and modality on the law side. In the present chapter we shall investigate the law-subject relation on the typical side of reality, and its bearing on the problem of individuality.

8.2 *Statistical measurements*
In experimental physics measurements are usually repeated many

23. Van Melsen A 138ff, C 148ff, 271ff. The quotation is from C 271 (my translation). The view that determinism is instrumental for any science is also expressed by Claude Bernard, cf. Kolakowski 90f.
24. Van Melsen C 300.
25. Van Melsen A 157ff, C 285ff.
26. Cp. Čapek 338ff.

times. Often every single measurement already yields a meaningful result, and one only repeats the measurement in order to improve on the accuracy by elimination of possible errors. In statistical measurements on the other hand, a single result has no immediate meaning. For instance, if we want to determine whether a certain die is a fair one, we have to perform a large number of trials to find out whether the distribution of throws over the six possibilities conforms to the law typical for a die.

Until the end of the 19th century this type of statistical measurements was not very important in physics. What is usually called statistical physics does not owe its name to its measurement procedures, but to a theoretical explanation (with statistical means) of macroscopic properties assumed to be generated as the average result of the relative motion and mutual interaction of the composing molecules. For this reason the theory of measurement as discussed in Chapter 3 may be called "classical".

Statistical measurements, especially those in (sub-) nuclear, atomic and molecular physics, first became important in the discovery of radioactivity. Their importance was enhanced in the interpretation of Brownian motion as measured by Perrin (1908) and the scattering experiments by Rutherford (1911).

Such an experiment may proceed in the following way. A number of atoms is prepared in the same initial state with the help of a so-called state selector. For instance, the atoms may all have the same initial momentum and energy. This state is disturbed by some interaction with a scattering system. Finally we measure how the state of the atoms is changed – for instance, we measure the angle of deflection. If the individuality structure of the incident atoms is known, we can determine something about the nature of the scattering system, or if the latter is known, we can learn something about the former. In this way Rutherford determined the size of the nucleus.

Generally we will find that all the atoms will not react in the same way to the disturbance. Therefore, we have to do this experiment with a large number of atoms in order to find the *spectrum* of the measurement results, and the *statistical distribution* of this spectrum. The former shows us the possible final states for the interaction; the latter is determined by the relative probability of a final state for a given initial state.

We can repeat the experiment for other initial states, and so experimentally determine the *transition probability*, connecting a certain initial state with a certain final state. These probabilities appear as the elements in Heisenberg's matrix representation of quantum mechanics.

Thus in statistical measurements we have both a *counting* procedure (the determination of the statistical distribution) and a *measuring* procedure (the determination of the spectrum of some measurable property). The latter does not differ basically from what we discussed in Chapter 3. Only if the spectrum is continuous it has to be broken up into a discrete number of intervals, in order to make it possible to count the number of occurrences in each interval. Counting is not directly possible with respect to a continuous spectrum.

In Sec. 3.7 we have seen that measurement, being essentially a comparison between what is simultaneous, always has a static character. That is the reason why the numerical and spatial modal aspects are also the most important aspects in the theory of statistical measurement. Classical probability theory is restricted to these aspects. The anticipations to the kinematic and physical aspects are not covered. Especially, the problem of the initial state cannot be solved other than by postulating its complete randomness, which is obviously difficult to reconcile with the alleged determinateness of motion and interaction according to classical physics. Probability as understood in quantum physics comes closer to a solution of these problems. We shall discuss this later on, but must first pay attention to the statistical distribution and the spectrum.

8.3 *Static theory of probability*
The theory of probability first of all has to account for two things: the spectrum of possible properties (which are simultaneously possible, so that the spectrum displays a spatial ordering), and the statistical distribution of relative frequency of occurrence of these possibilities, which has a numerical character. Formally, probability is defined as a numerical "measure" over a set of possibilities. In this section we shall briefly recall the formal topological properties of the spectrum and its measure.[27]

We call the set U of all possibilities the "universe of discourse", or "sample space", and we consider two operations on the sub-sets $A, B, \ldots$ of U:
– the union of two sub-sets: $A \cup B$
– the intersection of two sub-sets: $A \cap B$
An element of U is an element of $A \cup B$ if it is an element of A, or of B, or both. It is an element of $A \cap B$ if it is an element of both A and B.

We call A and B disjoint, if $A \cap B = \phi$, the empty set (con-

27. Nagel A 92ff; Bunge B 89-93; Popper A 326ff, C; Jauch; Hempel B 386ff; Hesse A, Ch. 5; Suppes A 274-291.

taining no element). A is a subset of B, or B includes A $(A \subset B)$, if $A \cup B = B$, or $A \cap B = A$. We call $-A$ the complement of A, if $(-A) \cup A = U$, and $(-A) \cap A = \phi$. $A \cup B$, $A \cap B$, and $-A$ are sub-sets of U, if A and B are.

The following set-properties can easily be derived:

$A \cap B = B \cap A$; $A \cap U = A$; $A \cup U = U$;
$A \cap A = A$; $A \cup B = B \cup A$; $A \cup A = A$;
$A \cup \phi = A$; $A \cap \phi = \phi$; $-U = \phi$; $-(-A) = A$.

These definitions and properties do not define a group, but a so-called Boolean algebra.[28]

Now, we define probability as a real numerical measure $P(A)$ for any sub-set A of U, such that:
- $P(A) \geq 0$
- $P(U) = 1$ (normalization)
- if $A \cap B = \phi$, then $P(A \cup B) = P(A) + P(B)$ – probability is an additive measure on the disjoint sub-sets of U.

This definition is sufficient to prove a number of theorems, such as:
- $P(\phi) = 0$
- $0 \leq P(A) \leq 1$, for any sub-set A
- $P(A \cup B) = P(A) + P(B) - P(A \cap B)$ for any two sub-sets A and B.

We define:
- conditional probability: if $P(B) \neq 0$, $P(A/B) = P(A \cap B)/P(B)$. Hence $P(A)$ is just short for $P(A/U)$.
- A is statistically independent of B if
  $P(A/B) = P(A)$ and $P(B/A) = P(B)$
  i.e., if they have "no common cause".[29]

This leads to the following theorems:
- if A and B are disjoint $(A \cap B = \phi)$: $P(A/B) = 0$
- if $A \subset B$: $P(A \cup B) = P(B)$, $P(A \cap B) = P(A)$, $P(A/B) = P(A)/P(B)$.
- if A and B are statistically independent: $P(A \cap B) = P(A).P(B)$.

Note that the property of statistical independence is a property of the spectrum and not of the statistical distribution. It is strongly connected with the concept of spatial independence.

Besides the probability function $P(A)$ we may have other numerical functions, e.g. $f(X)$ over the set U, where X denotes a sub-set of U. If we delimit the variable X to a set of non-overlapping (disjoint) sub-sets of U, such that their sum is U, we can define the mean value of $f(X)$ with respect to the probability measure $P(X)$ by:

28. Boole; Suppes A 202ff; another approach is that of a Borel set.
29. Reichenbach C 157ff.

$$< f(X) > = \sum_X f(X).P(X)$$

and the standard deviation $\Delta f(X)$ by:

$$(\Delta f(X))^2 = < f^2(X) > - < f(X) >^2$$

wherein $\Delta f(X)$ is a non-negative real number.

If U is a continuous set and x is a continuous variable defined in a certain interval $a \leq x \leq b$, the mean value of a function $f(x)$ is analogously defined as

$$< f(x) > = \int_a^b f(x).P(x)\mathrm{d}x$$

and, similarly, if U is a multidimensional continuous set. In this case, $P(x)$ is not a probability, but a probability density. $P(x_0)dx$ is now the probability that x has a value with the range $x_0 < x < x_0 + dx$. If the members of U are described by two variables x and y, we say that they are statistically independent, if $P(x, y) = P_1(x).P_2(y)$. In this case the possibilities in U can be ordered in a two-dimensional way characterized by the parameters x and y, which are now mutually orthogonal.

Probability as a measure on a set U is not the only measure satisfying the above-mentioned definitions and theorems. If U has a finite number u of elements, $P(A)$ can be interpreted as the number of elements in the sub-set A, divided by u, i.e., the relative number of elements in A. Since this interpretation has the same formal properties as probability the two are isomorphic. This isomorphy is the theoretical basis of the *measurement* of probability. It can also be used in the statistical definition of *entropy*.

If U is a spatial figure, and A is a spatial part of U, $P(A)$ can be interpreted as the spatial magnitude of A (i.e., its length, area, or volume) relative to that of U. This formal relationship with probability is used in the statistical conception of a phase space (see Secs. 6.6 and 8.6).

8.4 *Interpretations of probability*

The formal system described above does not determine the probability function beyond its limits. For all sub-sets A not equal to ϕ or U, $P(A)$ is only known to lie between zero and one. We need a further specification which can only be found by studying the typical properties represented by the set U. We recall that $P(A)$ is a "measure" or "weight function" of the sub-set A relative to U. We can consider three cases.

(*a*) It is often possible to assume on rational grounds that different sub-sets have equal weights, because of some symmetry relation. In this way we solve simple problems, such as occur in dice or card playing, assuming that the dice are not loaded and the card players are honest. In several more complicated problems which occur in quantum physics, for example, the symmetry of the systems concerned can facilitate their solution.

(b) Sometimes it is possible to design a theory to calculate weights which are not equal because of symmetry. In classical statistical mechanics one finds the beginning of this approach (cf. Sec. 8.5). It is fully developed in quantum physics (cf. Chapter 9).

(c) If there is no theory available, the only way to determine the probability function is by experiment. Even in this case the law is not reduced to the subject side. Also frequency hypotheses based on statistical extrapolation, such as mortality tables, can only be used if they are assumed to represent a law, since there is no logical justification for the conjecture that frequencies will remain constant, and thereby permit extrapolation.[30] Probably (without many exceptions), all statistics in the non-physical sciences is of this type. In the first two cases, (a) and (b), experiments also remain important, of course. Theories are never a priori, but hypothetical, and must therefore be checked experimentally.

The fact that the probability function depends on the typical structure represented by the set U, and that there are three possibilities of determining this function, has not always been recognized clearly enough. This may explain why there is so much disagreement about the interpretation of probability.[31] We shall briefly review some different views.

Keynes, Jeffreys and others,[32] who do not recognize the typical law determining the probability function and the set U, conceive of probability as a *logical* relation between propositions. The theory is especially designed to give account of the inference of laws from empirical facts.[33] In our view, the outcome of experiments as described in Sec. 8.2 reveals the individuality of the interacting subjects, and cannot be accounted for in a purely modal, let alone a purely logical, way. Probability does not describe our knowledge of physical systems, but their lawfully determined individual behaviour. Marge-

30. Popper A 168f.

31. Cf. Braithwaite; Carnap B; Jammer E 7; Nagel A; Margenau A, Ch. 13; Poincaré B, Ch. 11; Popper C.

32. Keynes; Jeffreys; Hesse A.

33. See Hempel B 57ff, 381ff, 385: A mathematical theory of "inductive probability" (as developed by Carnap) is only available for a relatively simple kind of formalized language; ". . . the extension of this approach to languages whose logical apparatus would be adequate for the formulation of advanced scientific theories is as yet an open problem".

nau rightly rejects this logicistic interpretation as being irrelevant in science.[34]

The "classical" interpretation, drafted by Pascal, De Moivre, Bernoulli, Laplace and others, is directed to the first possibility described above. It is applicable if the symmetry of the problem allows us to find disjoint sub-sets A of U, such that these sub-sets have equal weight and together add up to U.[35] Therefore, the classical interpretation assigns equal probabilities to "equally favourable cases". The founders of this theory were mainly inspired by games of chance, such as dice playing. This theory is also applied in statistical mechanics (cf. Sec. 8.5). It clearly breaks down if no equally favourable possibilities can be found.

The classical view is sometimes criticized because of its alleged circularity, the equally favourable cases being definable because they have equal probabilities. But, as our examples show, in those cases covered by the theory the equally favourable cases are inferred from the symmetry of the systems. Thus the classical theory can only be criticized because of its limited scope, and is, in fact, still of great importance – e.g., in quantum physics (see Chapter 9).

At the other extreme (clearly referring to the third possibility) one finds the definition of probability as a relative frequency of occurrence (Von Mises, Reichenbach).[36] As a definition, it reduces the law to the subject side, or metric to measurement. Indeed, the measurement of probability can only be performed by determining the relative frequency of the occurrences of every possible case.[37] But it seems a somewhat defeatist reaction to the failure of the classical definition if it assumes that in no case can one find a lawful metric for this probability. Anyhow, such laws can be found in quantum physics.

In his early publications Popper[38] defended a variant of the latter view. Later he developed the classical theory into the "propensity" interpretation of probability. It corresponds to the second possibility (b) mentioned above, but introduces "weighted" instead of "equal" probabilities. According to Popper, we have to ". . . interpret these weight of the possibilities (or of the possible cases) as *measures of the*

34. Margenau A 250ff; Popper C 29.
35. Popper A 168.
36. Von Mises 163-176; Reichenbach C 96ff.
37. Hempel B 387 essentially supports the view of Von Mises and Reichenbach, although he criticizes their formulations. They define probability as the limit of the relative frequency in an infinite series of performances, and Hempel rightly observes that such series are not realizable. But this criticism does not touch the heart of the problem – namely, that probability has both a law side and a subject side, and that the former cannot be reduced to the latter.
38. Popper A.

propensity, or tendency, of a possibility to realize itself upon repetition".[39]

Our view comes very close to Popper's interpretation as far as classical probability is concerned. We distinguish the formal theory described in Sec. 8.3 from the typical law which varies for different systems. We distinguish the law side, defining the set U of possible cases (the spectrum), and the probability function describing their weights, from the subject side (actual occurrences of the possible cases). We shall now discuss how these principles were applied in classical statistical mechanics, and then further investigate the relevance of the kinematic and the physical modal aspects for the theory of probability.

8.5 *Classical statistical mechanics*
The main application of probability theory in classical physics is statistical mechanics. It is based on the assumption that a gas, e.g., consists of a large number of similar molecules, which can only differ by their position, velocity, mass and moment of inertia. The kinds of motion considered are linear motion and rotation (sometimes also vibration). Rotation and vibration are only considered with respect to polyatomic molecules since atoms are supposed to be point-like. It is of interest to mention this because, if the finite extension of the atoms is taken into account, the method of classical statistical mechanics breaks down, as we shall see.

Statistical physics is often thought of as replacing thermodynamics, or at least providing its foundations. But statistics is not a purely modal theory (because of the assumption of the existence of similar molecules) and therefore cannot be a basis for the entirely modal thermodynamic theory. On the other hand the latter is inferior to statistical mechanics which can, by its nature, be applied to typical problems. Statistical mechanics is also easy to be incorporated into quantum physics – which is required if we wish to understand why classical statistical mechanics is applicable at all. Finally, as we saw in Chapter 5, thermodynamics is mainly retrocipatory, whereas statistical physics is anticipatory. In classical statistical mechanics there are two approaches, put forward mainly by Maxwell, Boltzmann, and Gibbs. We shall briefly discuss them in order to show the application of the formal theory as discussed above.

The starting point of the Maxwell/Boltzmann approach is the so-called *Maxwell distribution* for the molecules in an ideal gas.[40] The

39. Popper C 32, F.
40. Maxwell A; Born 50f.

following assumptions are made. (*a*) The molecules are fully described by their position **r**, velocity **v**, and mass m. (*b*) The particles do not interact with each other. This implies that the probability of finding a particle with a certain value for (**r**, **v**) is independent of the positions and velocities of other particles. Thus it is sufficient to derive the one-particle probability function which must be multiplied by the number (N) of molecules in order to find the distribution function for the gas. (*c*) The distributions for **r**, $f_1(\mathbf{r})$, and for **v**, $f_2(\mathbf{v})$, are independent: $f(\mathbf{r},\ \mathbf{v}) = f_1(\mathbf{r}).f_2(\mathbf{v})$. (*d*) There is equilibrium, which means that the distribution is spatially homogeneous (if there is no external field) and isotropic. The former means that $f_1(\mathbf{r})\mathbf{dr} = const.\ \mathbf{dr}$. The constant is found by normalization: it is equal to the inverse of the volume V of the gas. Hence, $f_1(\mathbf{r}) = 1/V$. The isotropy implies that $f_2(\mathbf{v}) = f_2(|\mathbf{v}|)$, or $f_2(\mathbf{v}) = f_2(|\mathbf{v}|^2) = f_2(v_x{}^2 + v_y{}^2 + v_z{}^2)$, i.e., the velocity is only dependent on the magnitude of **v**, and not on its direction. (*e*) The three coordinates v_x, v_y, v_z are independent, meaning that $f_2(\mathbf{v}) = f_x(v_x).f_y(v_y).f_z(v_z)$.

These five assumptions are sufficient to show that

$$f_2(\mathbf{v}) = a \exp -\tfrac{1}{2}m\beta\ (v_x{}^2 + v_y{}^2 + v_z{}^2)$$

The factor a can be found by normalization, the factor $-\tfrac{1}{2}m\beta$ by calculating the pressure P exerted by the gas on the wall. One finds the relation $1/\beta = PV/N$, which means that $\beta = 1/kT$ because of Boyle's law: $PV = NkT$ (T is the temperature, k is Boltzmann's constant, whose value is only determined by the choice of the units).

It will be clear that the Maxwell distribution is found from symmetry arguments. Several details can be criticized, and there are other derivations.[41]

Boltzmann recognized that the term in the exponent is just the kinetic energy of the molecule, divided by $-kT$. If we now introduce the concept of the state of the molecule, characterized by its velocity and position, we find that the relative probability of finding a particle in either one of two states with energies E_1 and E_2 is

$$P_1/P_2 = (\exp -E_1/kT)/(\exp -E_2/kT) = \exp -(E_1 - E_2)/kT$$

This was generalized by Boltzmann (and it is still the foundation of all statistical physics) to any system in equilibrium consisting of molecules or other particles which can freely exchange energy. It is nothing but an a priori assumption concerning "equally favourable cases". If two states have the same energy, their probability is the same. If they have different energies, their relative probability is given by the Boltzmann factor (as it is called). If the possible states have a continuous spectrum (which is the case for a classical gas), it is the probability density which is determined by the Boltzmann factor.

41. Born 51ff; Tolman, Ch. 4.

Whereas Boltzmann considered one system consisting of many molecules, Gibbs[42] studied an "ensemble", an infinite number of systems similar in their structure and boundary values, but with different "microstates". Above we defined the state of a *molecule* as being characterized by its position and velocity. The microstate of a *system* of molecules is the juxtaposition of all molecular states, while the macrostate of the system enumerates its macroscopically determinable properties, such as volume, pressure and temperature.

The microstate of a system can be represented by a point in a $6N$-dimensional phase space (N is still the number of molecules in the system). There is a many-to-one relationship between microstates and macrostates. Many microstates may correspond to a certain macrostate, but a microstate fully determines the corresponding macrostate. Therefore, if all microstates are equally probable, the relative probabilities of macrostates are proportional to the numbers of their corresponding microstates. In the case of a continuous spectrum of possibilities a macrostate can be represented by a region in the $6N$-dimensional space of microstates, and its probability is proportional to the volume of this region (cf. Sec. 6.6).

According to Gibbs all microstates are equally probable, as far as they are "accessible" by the system, i.e., as far as they are compatible with one or more restrictions or constraints. Thus in the case of a completely isolated system, for which the energy is constant, Gibbs introduces the "microcanonical ensemble". Here, all microstates with the same energy are equally probable (other states, not being accessible, have probability zero). For systems at constant temperature, for which the energy may fluctuate, one has the "canonical ensemble", in which the relative weight function for different microstates is the Boltzmann factor, $\exp -(E_1 - E_2)/kT$. Finally, if the number of molecules, as well as the energy, is undefined, one has the "grand canonical ensemble" with the weight function $\exp -(E_1 - E_2 - (N_1 - N_2)\mu)/kT$, called the Gibbs factor, wherein μ is the "chemical potential". In this theory, the entropy, the free energy, and other important thermodynamic variables can easily be defined.

The approaches of Maxwell, Boltzmann and Gibbs, are mentioned here, in the first place, to show the a priori character of their basic assumptions. These can only be justified ". . . by the correspondence between the conclusions which it permits and the regularities in the behaviour of actual systems which are empirically found."[43]

42. Cf. Kittel; Tolman 43ff.
43. Tolman 59; see also Kittel 34, 35; Popper A 208.

Our discussion also shows the typical and individual character of these theories. In both approaches, the *similarity* of the systems to be studied is a basic assumption. In the 19th century each molecule was assumed to be identified by its position and velocity at any time. But quantum physics has shown that the position of every individual molecule is not relevant, but only the distribution of all molecules over the accessible states. What is relevant is that a point in the six-dimensional phase space is occupied, not which molecule happens to be there.

The symmetry of the state function for similar particles allows two possibilities. A singular molecular state can be occupied by at most one particle if it is a fermion, or by an unlimited number of particles if they are bosons (cf. Ch. 10). The Maxwell-Boltzmann distribution is now only a limiting case (for very small occupation probabilities) of the more fundamental distribution functions: Fermi-Dirac, for fermions, and Bose-Einstein, for bosons. Whether a particle is a fermion or a boson is determined by its typical structure.

Once again this shows that statistical physics is not a fully modal theory, although it has very general features. In fact, the correct derivation of the classical Maxwell-Boltzmann distribution can only be given from quantum statistics because the classical assumption of complete identifiability of the molecules in kinetic terms leads to an overestimation of the number of possible microstates.[44]

There are more considerations to indicate that the classical approach can only be justified by quantum physics. One is the assumption that monatomic molecules have only three "degrees of freedom" (i.e., the number of coordinates necessary to specify the molecule's relative position), whereas diatomic molecules, for instance, have two additional degrees of freedom. (This refers to possible rotations about two independent axes. The relative vibration of the two atoms leads to another degree of freedom). Especially when the internal structure of atoms consisting of a nucleus and several electrons, was discovered, it became clear that this assumption is incomprehensible from a classical point of view.

However, quantum physics accounts for the existence of discrete energy levels which are dependent on the internal structure of atoms and molecules. These levels are widely spaced, such that electronic transitions from the ground state to the first or higher states do not occur at normal temperatures. But rotational states for diatomic molecules are less widely spaced, and therefore they can be excited easily at room temperature. This has consequences for the specific

44. See, e.g., Kittel 304-307, 390-392.

heat of a gas, which is nearly equal to $(3/2)NkT$ for a monatomic gas, as well as for a diatomic gas at 50 K, whereas this value increases to $(5/2)NkT$ for diatomic gases at higher temperatures. Both the temperature dependence of the specific heat, and more fundamentally, the applicability of classical statistics to normal gases, can therefore be understood only from the quantization of energy levels according to quantum theory.[45]

8.6 *Wave theory*

We have seen that probability is a measure over a set of possibilities. In a mathematical sense this is nothing but a field over a space or a region in a space, at least if the set of possibilities is continuous. Stating this once again shows the static character of classical probability theory, and points, at the same time, to the way to open it up in a kinematical sense. For as we saw in Chapter 7, the kinematics of a field leads to the theory of waves. This theory is not only applicable to physical fields in physical space, but to any field in any space, including probability.

We can also do the reverse. As soon as it was established, both experimentally and theoretically, that all particles move like wave packets, Born recognized these waves as probability waves. It should be emphasized that as long as we only consider the *kinematics* of kinematic subjects, we have no need of this interpretation. Therefore we discussed wave theory in a separate chapter, because it is of a purely modal character – in contrast to probability theory. The relation of wave theory and probability concerns the kinematic aspect of the latter, and manifests itself if we wish to account for the individuality of the physically qualified particles.[46]

Both theories have an anticipatory character. For the wave motion of physically qualified individual particles this implies that the probability interpretation becomes necessary as soon as we wish to consider the individual interaction of the particles with some other system (for example, but not necessarily, a measuring apparatus). A purely kinematic theory breaks down for this interaction.

The theory of waves as applicable to the kinematics of probability is not restricted to sinusoidal or spherical waves. It is the phase which shows the modal kinematic character of wave theory. The phase is not directly relevant for the probability, which is only determined by the amplitude. The most peculiar consequence of the

---

45. Cf. Mott.
46. Heisenberg's relations can now be interpreted as determining the minimum statistical spread or standard deviation in measurements.

phase formalism is that it leads to interference. Because probability is no longer static, but can be propagated, we have to reconsider the addition of probabilities.

In classical theory, addition of probabilities can only lead to an increase, because probability is essentially a positive measure. But the phase relations between interfering waves make it possible that waves may annihilate each other. In order to save the positive-definiteness of probability as a measure, it is defined as the square of the absolute value of the amplitude of the wave. (It is now also clear why wave packets must be normalized to unity, cf. Sec. 7.4). Especially this interference of probability, completely unknown in classical static probability theory, shows the relevance of the kinematic aspect of probability. It is one important new feature of quantum statistics.[47] A second innovation concerns the description of the transition from an initial state to a final state (cf. Sec. 9.8), which is also intimately related to wave theory.

8.7 *The physical qualification of probability*

Although probability is always presented as a numerical measure over a set of possibilities, it is always physically qualified in classical, as well as in quantum theories. In either case one of the more or less probable possibilities must be actualized. This actualization only occurs in some interaction – for example, shuffling cards, throwing dice, interactions in classical and quantum physics. The temporal order of possibility and its actualization is clearly asymmetrical, and therefore anticipates the physical modal time order of irreversibility. We shall make a few remarks about classical physics, before we start with the discussion of quantum physics in the next chapter.

First, we observe that the theories of Boltzmann and Gibbs lead to a description of the equilibrium state of a system as the most probable state. The irreversible approach to equilibrium is explained by the assumption that any actual system will "move" through all accessible states, such that the *spatial* average in phase space is equal to the *temporal* average for a single system. This so-called ergodic theorem (or a weaker "quasi-ergodic" theorem, according to which every accessible microstate will be approached arbitrarily close after some time) has been the subject of intensive mathematical research, but cannot be proved except for very simple systems under severe restrictions.[48]

47. Popper C underestimates this difference between classical and quantum statistics in his critique of the "great quantum muddle".
48. Truesdell 360-363; Khinchin, Ch. 3; Tolman 65ff; Penrose 39ff.

In our view, this problem cannot be solved in a purely modal theory, because it only has meaning if the systems in the spatial ensemble all have the same structure, whereas in calculating the temporal average it is assumed that the system retains its typical individuality during its passage through all possible states. The fact that the two averages must be the same is therefore not something which must be proved, but lies at the basis of all statistical methods. It assumes that the same system has a constant typical structure, or that similar systems are subjected to the same structural law. It says that the typical law is valid during any time, and for all systems under consideration.

Next we observe that the calculation of the entropy and related properties of a system is usually possible only for simplified systems of non-interacting molecules, such as the ideal gas, or the linear chain of magnetic molecules.[49] It is remarkable that such a system will not do the job. Because the molecules do not interact with each other, the microstate of the linear chain will never change, and in a perfect ideal gas mixture, there is no diffusion, e.g. If the microstate happens to correspond to a non-equilibrium macrostate (e.g., due to its preparation), it will never go to equilibrium as every actual system does. Thus we assume that there is some interaction between the molecules, small enough not to destroy the results of the calculation, but large enough to change the microstates so rapidly, that the temporal average may be equated with the calculated spatial phase average for the system.

Boltzmann systematically introduced the interaction in the six-dimensional phase space.[50] For an arbitrary (because unknown) interaction he substituted a collision probability function for pairs of molecules, describing the probability at any time for given positions and velocities before the collision, to find the change in velocity caused by the interaction. The theory leads to the so-called Boltzmann-equation which can account for many phenomena like viscosity, diffusion, thermal and electric conduction, if certain assumptions are made concerning the interaction. This approach is also useful in quantum physics. In this formalism a function H can be defined, which decreases in time for any system consisting of interacting molecules until the system has reached its equilibrium state in which H is constant. (This equilibrium state is again the Maxwell-Boltzmann distribution).[51]

49. Kittel, Ch. 2ff.
50. See, e.g, Tolman, Ch. 5 and 6.
51. As observed in Sec. 6.6, it is essential in this derivation that the state of a system be described by a domain (not a point) in state space.

166

The function H can be connected to the entropy of the system, and therefore Boltzmann's theory was hailed as deriving the irreversible approach to equilibrium from reversible kinematics. We have already discussed this in Chapter 6. Here it suffices to observe that this derivation depends essentially on the interaction between the molecules, and therefore is not of a purely kinematic character. Moreover, the derivation makes use of probability theory and must then distinguish between *actual* states in the past, and *possible* states in the future, meaning that irreversibility is presupposed from the start.

Because of the difficulties inherent in Boltzmann's and Gibbs' approaches as to the explanation of irreversibility, modern treatises no longer try to derive irreversibility from essentially mechanical systems. Rather, irreversibility is introduced from the outset. This is especially done in the form of Markov processes in which the state of a system is determined by the preceding states. This approach also has its difficulties, especially because of the continuity of time. On the other hand, however, it has possibilities not shown by the classical methods.[52]

Finally, we note that in all applications of probability theory the initial state forms a separate problem, at least in physics.[53] Although the initial state may be partly determined by some previous interaction or preparation, it necessarily has an amount of disorder, "molecular chaos", or "randomness".[54] One has tried but never succeeded in defining randomness. It appears that one has to accept it as a primitive concept. For instance, if we want to check the probability theory on dice playing, we assume that the way the dice are thrown does not influence the result in the mean. An honest card player is assumed to shuffle his cards at random. And in an opinion poll one has to strive for a representative sample. There are criteria to avoid biased samples, but there is no universal criterion to establish a completely random sample.

We consider randomness to be another expression of the individuality of the systems concerned, which cannot be fully delimited by specifying some of their properties. On the one hand, complete randomness does not exist. We can only make statistical predictions with respect to systems of which at least something is known of their typical structure. On the other hand, probability without randomness is useless. In quantum physics the initial state determining the statis-

52. See, e.g., Penrose.
53. In biology, psychology, or sociology, the related problem is that of the "population", the "Kollektiv", or a "representative sample".
54. Hempel B 386; Nagel A 32ff; Popper A 151ff, 359ff.

tical distribution also contains an element of randomness. According to a theorem related to the Heisenberg relations, if any property of a system is completely determined by its preparation, the "canonically conjugate" property is completely random. In general, the initial state in quantum physics can be better specified than in classical physics. But even then it always contains an undetermined phase.

9. Probability in Quantum Physics

9.1 *Stating the problem*

The development of the basic concepts of quantum physics involved a good number of years of concerted efforts on the part of many theoretical and experimental physicists in many countries. The basic ideas of the theory were essentially established during a thirty year period (1900-1930), yet even now there is still no agreement about the interpretation of its foundations. In this chapter it is not our intention to present the historical development of the theory, but rather to discuss the mathematical framework, which was basically established in the years 1925-1930 by physicists such as Schrödinger, Heisenberg, Born, Jordan and Dirac, and by mathematicians such as Von Neumann.[1] We will restrict ourselves to the so-called Hilbert-space representation (Sec. 9.2), which, though not the only one[2] nor the most general one, nevertheless suffices for our purpose. Our purpose is, namely, to show the probabilistic character of quantum physics, and to point out how it differs with classical probability theory.

The general theory of quantum physics addresses five related problems (cp. Chapter 8):

(*a*) To find the spectrum of possible properties of the system under study (Sec. 9.3).

(*b*) To give an objective description of the (initial) state of the system, to the extent that it is specified, and to the extent that it is at random (Sec. 9.4).

(*c*) To determine the relative statistical weights associated with the possible properties of the system relative to its state. In this context we have to discuss the external (modal) symmetries (Secs. 9.5, 9.6) as well as the internal structure, partly expressed by internal symmetries (Sec. 9.7).

(*d*) To determine the temporal development of the state during the time from one interaction to the next, and to treat the problem of interference (Sec. 9.8).

---

1. Cf. Jammer D 307ff.
2. Jammer D 315.

(e) To explain the actualization of one of the possible properties via an interaction, which implies the distinction between "possessed" and "latent" properties (Sec. 9.9).

It is most remarkable that at least the first four problems can be treated within the context of a single concept, that of a complex Hilbert space, with its associated hermitean operators. This concept is an abstract one, and has many realizations. All Hilbert spaces with the same number of dimensions are isomorphic to each other.

The basic hypothesis of quantum physics says that the set of possible *states* of a system is isomorphic to all Hilbert spaces of a certain dimensionality, which depends on the typical structure of the system.[3]

Any *property* of the system is related to a coordinate system (a set of basis functions) in Hilbert space, such that the property's spectrum is related to the dimension of that space. Properties with a number of possible values less than the dimension of the Hilbert space are called "degenerate" for that system. Degeneracy is always connected to some kind of symmetry.

The *probability* associated with a certain value of some property is determined jointly by the spectrum of that property and the state at the moment the interaction revealing (probing) that property takes place. Thus, while the concept of a Hilbert space provides the description of probabilities in an isolated system, at the same time it anticipates interaction. Its use breaks down as soon as we want to investigate the interaction itself. So the fifth problem mentioned above is at best only partly solved.

9.2 *Operators*

In Sec. 2.4 we mentioned the properties of a Hilbert space: the dense and complete set of all linear combinations of a number of basis functions with complex coefficients. In this space for any pair of functions f_1 and f_2, a linear functional (f_1, f_2) exists and is called the scalar product. Now the concept of a linear operator is introduced as a "mapping" of the Hilbert space onto itself, more or less similar to a rotation in an Euclidean space of two or three dimensions.[4]

3. For the sake of argument, we shall assume that the number of dimensions of Hilbert space is denumerable.

4. This means that an operator in the Hilbert space manifests itself as a "matrix" by which the components of a function in the space are transformed.

If f and g are arbitrary functions in the Hilbert space H, then A is a linear operator if it transforms the function f into Af such that:

- Af is a function in H
- if p is a complex number, $Apf = pAf$
- $A(f + g) = Af + Ag$ (A is linear)
- if besides A, also B is a linear operator in H, then
  $(A + B)f = Af + Bf$
  $A(Bf) = (AB)f = ABf$
- if $Af = 0$ for any f, A is the "zero operator"
- if $If = f$ for any f, I is the "identity operator";
  for any operator, $IA = AI = A$
- to A corresponds an "adjoint operator" A^+, such that
  $(f, Ag) = (A^+f, g)$; this implies that $I^+ = I$; in matrix notation, $(A^+)_{kl} = A_{lk}{}^*$, where the star denotes complex conjuga-
  tion.

Addition of two operators is commutative ($A + B = B + A$) and all linear operators in a Hilbert space form a group with respect to addition, with the zero operator as identity element, and with $-A = (-1)A$ as the inverse of A. In general, multiplication of operators is not commutative. We say that A and B commute if $AB = BA$.

In quantum physics one is especially interested in so-called hermitean operators as well as in unitary operators.

A *hermitean* or *self-adjoint* operator is defined by the property $A = A^+$, i.e., for any pair of functions f and g in H, $(f, Ag) = (Af, g)$. Each hermitean operator generates a basis in the Hilbert space. This means, for any hermitean operator A there exist vectors n_i such that $An_i = a_i n_i$, where a_i is a *real* number. If normalized, the so-called *eigenvectors* or eigenfunctions n_i of A have the properties of basis functions in H: $(n_i, n_i) = 1, (n_i, n_j) = 0$, for any i and j, $i \neq j$. Moreover, the set of n_i's is complete, which means any function in the Hilbert space can be written as a linear combination of those eigenvectors.

The real *eigenvalues* a_i can serve to distinguish the eigenvectors. Therefore, if two mutually orthogonal eigenvectors have the same eigenvalue, all vectors in the two-dimensional space consisting of the linear combinations of these two eigenvectors are also eigenvectors. Hence an eigenvalue determines a subspace of Hilbert space, whether one-dimensional (non-degenerate eigenvalue) or multi-dimensional (degenerate eigenvalue).

If two hermitean operators *commute*, they have the same set

171

of eigenvectors, but with different eigenvalues, and different degeneracies. To every unit vector n_i of a basis in Hilbert space is connected a hermitean projection operator P_i, which transforms any function into its projection onto that unit vector. Thus, because $f = \Sigma(f, n_i)n_i$, we have $P_i f = (f, n_i)n_i$. For the basis vectors themselves, $P_i n_i = 1$ and $P_i n_j = 0$ if $i \neq j$. Hence the eigenvalues of P_i are either one or zero (the latter is highly degenerate), and the projection operators can be used to describe yes-no experiments.[5]

A *unitary* operator U is defined by the property $UU^+ = I$, the identity operator. Thus unitary operators can form a multiplication group with I as the identity element, and U^+ as the inverse of U. The application of a unitary operator to a basis leads to a new basis having again the orthogonality and normalization properties. With this change of basis, a hermitean operator A is transformed into U^+AU. If A and U commute, A is not changed ($U^+AU = U^+UA = A$). Therefore, unitary operators are very useful in describing symmetry operations, in which transformations of the state of the system are made without changing its properties.

Often a unitary operator can be written as $U = \exp iS$, where S is a hermitean operator, called the *generator* of U. If U acts on an eigenvector n of S with eigenvalue q ($Sn = qn$), we find:

$$Un = (\exp iS)n = \exp i(Sn) = \exp i(qn) = (\exp iq)n$$

Since q is a real number, $\exp iq$ is a complex number with absolute value 1 (cf. Sec. 2.4). In this particular case, if we apply the unitary operator to this eigenvector of its generator, we find that the eigenvector is merely multiplied by a complex number. This is especially relevant if all eigenvalues of S are the same (complete degeneracy) for a certain system, in which case any function in the Hilbert space is an eigenvector of S, and the unitary operator does nothing but multiply all functions in Hilbert space with the same complex number. This leaves invariant the scalar products of all pairs of functions in this space.

---

5. Cp. Jauch 73: ". . . every measurement on a physical system can be reduced, at least in principle, to the measurements with a certain number of yes-no experiments". It should be realized that a yes-no experiment is a theoretical idealization. In principle, no actual physical measuring instrument can be reduced to a perfect yes-no experiment because of noise. Just because every physical instrument is a quantum mechanical system itself, this noise cannot be made as small as one likes. This refutes the possibility of accepting yes-no experiments as an *empirical* basis for quantum physics, although they may be used as a *theoretical* starting point.

This procedure turns out to be particularly useful for the description of the spatial and temporal homogeneity for isolated systems (Sec. 9.5). For spatial isotropy, one finds that the degeneracy of eigenvalues is not complete, so that the corresponding unitary operator is a two- or more-dimensional matrix (Sec. 9.6).

9.3 *Static probability theory*
Thus far we have merely described a formal, mathematical scheme. Now, in order to discuss its physical interpretation, we have to state in which way this formalism is an objective representation of physical states of affairs. Such a statement has the character of a hypothesis, which can never be proved analytically, but must be corroborated by its consistency with other theories and experimental results. For the time being, we shall only deal with the first three problems mentioned in Sec. 9.1. We first investigate the numerical and static aspects of probability in quantum physics (compare Sec. 8.3).

On page 60 we defined a *magnitude* as a property of comparable subjects, which allows a quasi-serial ordering of those subjects. Now one asserts that any physical magnitude 'A' for a system is represented by a hermitean operator A in a Hilbert space. The eigenvalues of A (whether discrete, denumerable, or continuous) compose the spectrum of possible real values for the property 'A' for that particular system – i.e., the spectrum is in part determined by the typical structure of the system. This interpretation rests on the fact that the basis functions of any orthogonal basis in a linear function space can always be quasi-serially ordered.

The *state* of an isolated system corresponds to a normalized function or vector in the Hilbert space. This is a somewhat idealized statement, because usually the state of a system must be represented by a "mixture", a statistical ensemble, of such states. Such an ensemble can be described with the help of a so-called "density operator" (cf. Sec. 9.8). For simplicity we restrict our discussion most of the time to "pure states", which can be represented by vectors in Hilbert space. The theory can be extended to include non-isolated systems, if the action from the outside is static (e.g., an electric or magnetic field), and if there is no reaction of the system to the outside world.[6]

If f represents the state of the system, the *statistical distribution* over the spectrum of the physical magnitude 'A' is determined by the scalar product of f and the eigenvectors of the operator A. Therefore, for an eigenvector n_i of A, whose eigenvalue is a_i, $(n_i, f)(f, n_i)$

---

6. Jauch 157ff.

is the probability that the subject will adopt the value a_i for A by virtue of an interaction in which the property 'A' is displayed. This applies in particular, but not exclusively, to a measurement of the magnitude 'A'.

The state functions f are restricted to normalized vectors in Hilbert space because the probability must be normalized. This can be seen because $\Sigma(n_j, f)(f, n_j) = (f, f)$ includes all possibilities for 'A', and must therefore be one. This means that if f itself happens to be an eigenvector of A with eigenvalue a_f, then the probability of finding a_f for the property 'A' is unity, and the probability of finding any other value for 'A' is zero. (This "certainty" in finding a particular value for a property 'A' is only possible for the idealized pure states). The mean value for the property 'A' with respect to a given initial state f is simply (f, Af).

Thus far the static theory already shows some remarkable differences from the classical theory discussed in Sec. 8.3. In the first place, we note that the spectrum of possible values for a property of a system reflects more directly its spatial character in that it is represented by a coordinate system, a basis in Hilbert space.[7] It still refers back to the numerical modal aspect, as is fitting for a magnitude.[8] The spectrum of possibilities, being determined, as it is, by the internal typical structure of the system, is shifted therefore to the law side. Hence the spectrum, rather than being primarily a subject-subject relation, determines the latter. Consequently, 'A' is no longer a *property*. No longer does it belong to the subject, but rather to the typical law to which the subject responds. This is an important point to keep in mind in our forthcoming discussion.

Next we note that also the concept of probability itself, as used here, differs from the classical one. In the latter theory probability is a *function* over the possibilities. In quantum physics, probability is a *functional*, a scalar product between the state function f and the eigenvector corresponding to a particular possible case. Thus we find that in quantum physics probability is a relation between the state of a system and the state of some reference system.

Actually, in classical theory as well, the probability is determined by both the spectrum of possibilities and the initial state, but the latter is assumed to be completely at random. In this respect the new theory is much richer than the old one, because it allows specification of the initial state. Now we have precisely indicated how the prob-

7. It is no longer a Boolean algebra; cf. Weizsäcker B 237.
8. Cf. page 47.

174

ability is jointly a numerical measure of the state of the system and the reference system.

We have rather vaguely spoken about ". . . the probability that in some interaction the property '*A*' is displayed . . ." and about a "reference system". We defer until later a more intense discussion of these concepts. We only stress now the *relational* character of the state. It only has meaning if referred to some property, which, however, can manifest itself only in an interaction.

Following the discovery that spatial position and kinematic motion are subject-subject *relations*, relative to spatial or kinematic reference systems, we find now that the physical state of a system can only be understood relative to a physical reference system. Such a system is objectively represented by a basis in Hilbert space, but is in any actual interaction a second physical system, with which the former system may interact.

This second system may be a measuring instrument, in the same way spatial coordinate systems and kinematic reference systems may be constructed from concrete metre sticks and clocks. But in an abstract analysis of physical interaction it is not necessary to restrict oneself to measuring instruments, neither in geometry and kinematics, nor in physics. On the contrary, in each of these three modal aspects the relation between a subject and a reference system or measuring instrument is merely one of many concretizations of an abstract relation, the modal subject-subject relation. Indeed, for the spatial, kinematical, and physical modal aspects we have found the modal subject-subject relations to be, respectively, relative position, relative motion, and relative interaction. Therefore we are justified in concluding that the theory has a relational character.

9.4 *State preparation, randomness, and complementarity*

The meaning of a state as represented by the state function *f* is given in its relation to a reference system representing a possible second system, with which it may interact. The state itself is the result of a previous interaction, and is intermediary between the two interactions. In experimental physics one speaks of a state selector as a device which e.g. singles out from an incoming beam of particles a certain state, excluding all other states. Hence it is a yes-no experiment, which can be described with a projection operator. Alternatively, a state selector is characterized by an eigenvector of an operator representing the property according to which the incoming particles are discriminated. The state of the particles emerging from a pure-state selector is independent of the state of the incoming particles, whereas the effect of a mixed-state selector depends partly on this preceding state.[9]

Restricting ourselves to a pure non-degenerate state selector (which gives the maximum obtainable determination of a state) we find that this state represents randomness in two ways. The first is connected with the phase of the state as follows. We have seen how the scalar product (f, n_i) determines the probability that a system in the state f will adopt the eigenvalue a_i as a result of an interaction characterized by the corresponding operator A. In particular, these probabilities are invariant under multiplication of the state vector by a complex number $\exp iq$, where q is a real number called the "phase" of the exponential function. Therefore, in state selection, the phase associated with the state f is completely undetermined, and must be assumed to be a random parameter.[10]

Next we recall that the state f produced by a state selector is an eigenvector of the operator A corresponding with the property 'A' according to which the systems are selected. This means that for a later interaction characterized by an operator B, f is not an eigenstate of B unless A and B commute. Thus, e.g., if A is the momentum operator, and B the position operator (which operators do not commute), then the particles emerging from the state selector may be said to have a precisely determined momentum, but not a precise position at any time before the next interaction takes place.

However, this statement is liable to be misunderstood, as the history of quantum physics has shown. The properties of physically qualified subjects as represented by hermitean operators have the character of a *law*, and therefore never *belong* to the subject. They display themselves only if the system interacts with some other subject. As long as we talk about an isolated system in the state f, we have to assume that any property of the system has a potential character, even if f happens to be a pure eigenstate of some operator. This latter case must be understood as a limiting case, and as such is comparable with the state of rest in kinematics and a one-dimensional "space" in geometry. As long as the system does not interact, the state is more or less autonomous with respect to any property. Indeed, the state has a "potential"[11] or "latent"[12] character.

In this respect the quantum physical concept of a state is profoundly different from the classical molecular state, or the micro- or

9. Houtappel et al. 611.
10. Tolman 349ff.
11. Heisenberg C 38.
12. Margenau A 175, 335.

macrostates of macroscopic systems (cf. Sec. 8.5).[13] The classical state is a mere enumeration of actual properties of the system, whereas the state in quantum physics has a potential character for a single system. For an ensemble of similar systems in the same state, it determines their properties in the mean.[14] It must be emphasized that this potential character of the state strictly pertains to isolated systems only and is therefore somewhat abstract. Any concrete system always interacts with other systems in various ways, and therefore its state is always actualized in some sense or another.

The incompatibility of properties represented by non-commuting operators has given rise to the concept of *complementarity*. It originated in the so-called Copenhagen School, mainly represented by Niels Bohr, and strongly bears the influence of some widely differing philosophies.[15] As a result, every proponent of complementarity has his own interpretation of it. One difficulty in understanding this concept is that it was introduced from the very beginning of the development of quantum physics. Therefore, its meaning has evolved along with the theory itself.

For some people complementarity is the same as the wave-particle duality. For others it refers to the relation between the microsystems governed by quantum physics and the macroscopic measuring instruments which are supposed to be describable in classical terms.[16] Sometimes it even denotes the psychic subject-object relation between the observer and the observed system. Shortly before the final establishment of Schrödinger's wave theory and Heisenberg's matrix mechanics in 1925, Bohr, Kramers and others paid much attention to the "complementary" relation between a "causal" and a "spatiotemporal" description of physical processes. In its simplest form the principle of complementarity merely expresses the fact, mentioned above, that an eigenvector f of an operator A cannot be an eigenvector of a different operator B, unless A and B commute.[17]

The fact that the position and momentum operators do not commute has particularly been the cause of much discussion. In the first place, it refutes the classical maxim that the state of any physical

13. Jauch 90ff.
14. Bunge B 249.
15. On the historical development of the concept of complementarity, see Jammer D 345ff, E; on the views of Bohr, see Bohr A, B, C, D; Meyer-Abich, Ch. 3; Bunge I 173-209; Feyerabend C; Hanson A; Holton A 115-161; Hooker.
16. Bohr A 209, 210.
17. From a historical point of view, this interpretation rests on a misunderstanding (first by Pauli) of Bohr's ideas; cf. Meyer-Abich 152.

system must be describable in modal kinematic terms, i.e., by its *actual* position and momentum at a given time. In the Copenhagen School, in an effort to focus on the classical maxim, attempts have sometimes been made to reduce all types of complementarity to this one of position and momentum (considered as "primitive"). But other kinds of incompatible magnitudes cannot be reduced in this way (cf. Sec. 9.6), and this idea had to be abandoned.

The need for a kinematic description disappears as soon as one recognizes the mutual irreducibility of the physical and kinematical modal aspects, which implies that the state of a physically qualified system must be referred to a *physical* coordinate system and not to a kinematical one. The situation is quite analogous to Einstein's criticism of the classical theory of motion, where in discussing kinematical temporal relations, we have to use kinematic frames of reference, in which, e.g., static simultaneity is relativized. Similarly, if we want to study the relations between physically interacting systems, we have to refer to physical frames of reference. This principle regarding physical frames of reference has consequences even for temporal relations which are not original to the physical modal aspect, such as position and momentum.[18]

Thus the potential character of the state must be discussed taking into account all three basic distinctions of our philosophical framework. Distinguishing law side and subject side, we discover that the properties of a system have the character of a law, and cannot be possessed by the system, apart from its interaction with other systems. Distinguishing modality and typical individuality, we allow for the contingency of the state concurrently with the fact that the main properties in which one is interested (position, momentum, energy) possess a modal, universal nature. Differently structured systems can interact just because they have such universal properties in common. Finally, the distinction of the modal aspects helps us to understand the relativization of the pre-physical modal relations in physically qualified relations.

9.5 *Modal symmetry: energy and momentum*
We proceed now to consider the problem of symmetry in more detail. Symmetry can be divided into two types, modal or external, and typical or internal. In both cases we can translate the problem of symmetry into spatial terms. Namely, we inquire as to which transformations in Hilbert space leave both the statistical distribution over some hermitean operator and its eigenvalue spectrum unchanged. All transformations of this kind are represented by unitary

18. Jauch 69, 112ff.

operators (which commute with the hermitean operator being considered). Further, with any kind of symmetry, there corresponds a *group* of unitary operators. Thus the theory of groups is of great use for the solution of symmetry problems.

The case of modal symmetries involves transformations which depend on the mutual irreducibility of the first four modal aspects, whereas the case of typical symmetry depends on the typical structure of the system concerned, and is therefore of a less general character (cf. Sec. 9.7). The modal symmetries depend on the isotropy and homogeneity of time and space, and the Galilean or Lorentz invariance of uniformly moving physical systems. Since we shall be concerned with isolated systems, we postpone the discussion of the internal time evolution until Sec. 9.8, and start with stationary states. This leads to the conservation laws of energy and momentum (the present section) and of angular momentum (Sec. 9.6). These laws were discovered prior to the quantum era, yet are far easier to derive in the quantum theory than in classical physics.

A unitary operator can be described by a matrix which can be broken down into a number of matrices of smaller dimensionality. The simplest case is that of a matrix of dimension 1 whose element is a single complex number of absolute value 1. This one-dimensional matrix is nothing but the exponential function exp iq, the "phase" q being some real number. This is the so-called phase-factor.

If we multiply all functions in Hilbert space with the same phase factor, then all scalar products (f, n_i), as well as all mean values (f, Af) for any linear hermitean operator A, are invariant under such a transformation. This phase-factor invariance allows us to study the conservation laws of energy and momentum for stationary states.

Consider an isolated system whose state at a certain time t is denoted by the function $f(t)$. We suppose that the state $f(t_1)$ evolves continuously from the state $f(t_0)$. Invoking Taylor's theorem, we find

$$f(t_1) = f(t_0) + (t_1 - t_0)\frac{\delta f(t_0)}{\delta t} + \tfrac{1}{2}(t_1 - t_0)^2\frac{\delta^2 f(t_0)}{\delta t^2} + \cdots$$

which can formally be written as

$$f(t_1) = \left[1 + (t_1 - t_0)\frac{\delta}{\delta t} + \cdots\right]f(t_0)$$

$$= \left[\exp i(t_1 - t_0)\frac{1}{i}\frac{\delta}{\delta t}\right]f(t_0) = [\exp i(t_1 - t_0)H]f(t_0)$$

179

where we have introduced the hermitean operator $H = \frac{1}{i}\frac{\delta}{\delta t}$ as the generator of the unitary operators describing the temporal homogeneity.[19]

Now we demand that the states $f(t_1)$ and $f(t_0)$ differ only by a phase factor. Because of the previous result, this phase factor can only have the form of $\exp i\omega(t_1 - t_0)$, where ω is some real number. That is,

$$f(t_1) = [\exp i\omega(t_1 - t_0)]f(t_0)$$

or: $(\exp -i\omega t_1)f(t_1) = (\exp -i\omega t_0)f(t_0) = g(\omega)$

where $g(\omega)$ is a Hilbert space vector independent of time, and characterized by the real number ω. In terms of $g(\omega)$, the general state vector is

$$f(t, \omega) = (\exp i\omega t)g(\omega)$$

If we apply the previously mentioned hermitean operator H (the "Hamiltonian") to f, we have

$$Hf(t, \omega) = \frac{1}{i}\frac{\delta}{\delta t}(\exp i\omega t)g(\omega) = \omega f(t, \omega)$$

Thus the real number ω is an eigenvalue of H. Since ω can be any real number, the spectrum of H is continuous, and consists of all positive and negative real numbers. The value of ω depends on the reference system, and may therefore be determined by the state selector. If the latter is incompatible with H or if it only produces mixed states, the state of the system corresponds with a statistical distribution over the spectrum of H. In the pure-state case, ω is the same for all vectors in the Hilbert space representing the internal structure of the system. This space is therefore an invariant sub-space of a much "larger" Hilbert space (of higher dimensionality) which includes all possible values for ω.

In a completely analogous way, we can consider the system at different positions $\mathbf{r}$ (or relative to different spatial coordinate systems). Then we find

$$f(\mathbf{r}_1) = [\exp i(\mathbf{r}_1 - \mathbf{r}_0).\mathbf{P}]f(\mathbf{r}_0) = [\exp i\mathbf{k}.(\mathbf{r}_1 - \mathbf{r}_0)]f(\mathbf{r}_0)$$

where $\mathbf{P} = \frac{1}{i}\nabla = \frac{1}{i}\left(\frac{\delta}{\delta x}, \frac{\delta}{\delta y}, \frac{\delta}{\delta z}\right)$

If $g(\mathbf{k})$ is a vector independent of position, characterized by the three-dimensional vector $\mathbf{k}$, then the general state vector can be represented by

$$f(\mathbf{r}, \mathbf{k}) = (\exp i\mathbf{k}.\mathbf{r})g(\mathbf{k})$$

19. This procedure was invented long before the rise of quantum physics, by Lagrange; cf. Jammer D 224.

P is again a hermitean operator with eigenvalues **k**, whose spectrum covers triplets of all possible positive and negative real numbers. It is the generator of the unitary operators exp *i***r**.**P**, which form a multiplication group isomorphic to the three-dimensional addition group of vectors **k**.

So far we have only studied the temporal and spatial homogeneity of the systems, referred to different temporal and spatial frames. In order to include uniform motion, we first connect the two cases to obtain

$$f(t, \omega; \mathbf{r}, \mathbf{k}) = [\exp i(\omega t + \mathbf{k}.\mathbf{r})]g(\omega, \mathbf{k})$$

This result is just the plane wave discussed in Chapter 7. Thus in a very natural way we arrive at the plane wave representation of temporal and spatial homogeneity, which in classical physics can only be found in a very laborious way. In Chapter 7 we saw that the number ω has the character of a *frequency* and is proportional to energy, and that the vector **k** is the *wave vector* proportional to the linear momentum. That is to say that ω and **k** are "constants of the motion". By considering Galilean transformations between inertial reference systems moving relative to each other, we find as in classical physics[20] $\omega = k^2/2m$, where k is the absolute value of **k**, and m is a frame-independent magnitude, called the mass of the system. Therefore, we find that the system satisfies the Schrödinger equation

$$Hf(t, \omega; \mathbf{r}, \mathbf{k}) = \frac{\mathbf{P}^2}{2m} f(t, \omega; \mathbf{r}, \mathbf{k})$$

If we consider the Lorentz instead of the Galilean transformations, the relation between energy E and momentum p becomes $E^2 = m^2c^4 + p^2c^2$. This means that for a certain value of the momentum (or wave vector) there are two possible values of the energy (or frequency), one positive and one negative. This was first pointed out by Dirac, and because this relation is universally valid, each particle has a corresponding "antiparticle", whose energy, or alternatively, charge is of opposite sign to that of its counterpart. Thus the positively charged positron corresponds to the negatively charged electron, and the negatively charged antiproton corresponds to the positively charged proton. Additionally, as regards some properties of nuclear interactions, a particle is the antipode of its antiparticle, and vice versa. But essentially they have the same mass.

In general, a moving system will not be represented by a single vector, but by a mixture or wave packet

20. Jauch 198; Kaempffer, Ch. 9.

$$h(\mathbf{r},\ t) = \int_{-\infty}^{\infty} A(\mathbf{k})[\exp i(\mathbf{k}.\mathbf{r} + \omega t + \varphi)]g(\omega, \mathbf{k})d\mathbf{k}$$

where the amplitudes $A(\mathbf{k})$ and the phases φ are determined by the state selector.[21] Hence we find that the absolute value of $h(\mathbf{r}, t)\mathrm{d}\mathbf{r}\mathrm{d}t$ is the probability that during the temporal interval between t and $t + \mathrm{d}t$, the system will be in the spatial interval between $\mathbf{r}$ and $\mathbf{r} + \mathrm{d}\mathbf{r}$. Thus the amplitudes $A(\mathbf{k})$ are subjected to the normalization condition

$$\int_{-\infty}^{\infty} h(\mathbf{r}, t)h^*(\mathbf{r}, t)\mathrm{d}\mathbf{r}\mathrm{d}t = 1$$

In a similar way, we can express the probability that the energy and the linear momentum have values in a certain interval. The general theory of the Hilbert space formalism now leads to the same Heisenberg relations between the minimal spreads in the statistical distributions of energy/momentum and time/position, as can be found from simple wave theory (cf. Chapter 7). We shall return to the kinematics of isolated systems in Sec. 9.8.

9.6 *Spin*

A somewhat more complicated problem is the one concerned with the rotational invariance of isotropic space. In classical physics it had already been shown that this symmetry leads to the conservation of angular momentum. A group-theoretical analysis rules out a description of this symmetry in terms of one-dimensional matrices, which precludes the phase factor formalism.

At some stage of the development of quantum physics it was assumed that all relevant operators could be derived from those of energy, position and momentum, just as in classical physics.[22] One problem in the application of this so-called Correspondence Principle[23] is that the momentum and position operators do not commute with each other, which has no analogy in classical physics.[24] But even if this difficulty is circumvented, one arrives at incorrect or incomplete results, as in the case of angular momentum.

In the ordinary representation of classical physics for simple

21. A pure state cannot be normalized, and is therefore unsuited for the representation of a physical system.

22. This view is not justified, not even for classical physics: the formulation as given in the text is a simplification.

23. See for the Correspondence Principle, Jammer D 109-118; Meyer-Abich; Hanson A, B, D 60-70; Messiah 29-31; Heisenberg C, Ch. 3; Petersen.

24. Margenau, Cohen 85, 86. For this section, see e.g. Messiah Ch. 3 and 13, p. 195ff, 508ff; Jauch, Ch. 14; Kaempffer, Ch. 1, 12.

systems (consisting of mass points), the angular momentum is $\mathbf{l} = \mathbf{r} \times \mathbf{p}$, where $\mathbf{r}$ is the position vector, $\mathbf{p}$ the linear momentum, and $\mathbf{r} \times \mathbf{p}$ is a vector product. Since the position operators involve simple multiplication by x, y and z, we can (according to the Correspondence Principle) derive the operator

$$l_z = \frac{1}{i}\left(x\frac{\delta}{\delta y} - y\frac{\delta}{\delta x} \right)$$

for the z-component of the vector $\mathbf{1}$. The corresponding eigenvalue equation can be solved. The eigenvalues are just the integers, $m = \pm 1, \pm 2$, etc. This solution is wrong, in so far as it gives only integral values for m, whereas experiments show that m can also have half-integral values, $m = \pm\frac{1}{2}, \pm\frac{3}{2}$, etc. It is impossible to find these half-integral values by merely looking for an analogy in classical mechanics.

We can arrive at the correct solution in two ways. First we observe that the angular momentum components, l_x, l_y, l_z as defined above satisfy the commutation relations

$$l_x l_y - l_y l_x = il_z \text{ (cyclic in } x, y \text{ and } z)$$

We then show that with this relation not as a result, but as a starting point, we find both integral and half-integral eigenvalues.

Alternatively, and without reference to classical physics, we can use group theory, and show that the unitary operators describing rotational invariance are reducible to matrices with one or more dimensions. Matrices of even dimensionality correspond to half-integral values for the spin, and those of odd dimensionality to integral values. If the dimensionality is n, the eigenvalues are: $-\frac{1}{2}(n - 1), \ldots, +\frac{1}{2}(n - 1)$, at unit intervals, such that the total number of different eigenvalues is just equal to n. The number n depends on the structure of the system, and is an invariant. The number $\frac{1}{2}(n - 1)$, the highest eigenvalue for each spin component, is also an eigenvalue of the total angular momentum operator $\mathbf{J}$, which must be sharply distinguished from the three components, l_x, l_y, l_z.

Only those eigenstates of the system can be superposed which have the same eigenvalue of $\mathbf{J}$. This is a so-called super-selection rule, to which we shall return later. On the other hand, for example, the superposition of eigenstates with different eigenvalues for the operator l_z is always possible. Thus for a spin-zero particle (e.g., a pion), $n = 1$, $J = 0$, and $m = 0$. For a spin one-half particle (e.g., an electron or a proton), $n = 2$, $J = \frac{1}{2}$, and m can have two values: $\pm\frac{1}{2}$; and for a spin-one particle (e.g., a photon), $n = 3$, $J = 1$, and $m = \pm 1$ or 0. More complicated systems like nuclei and atoms may have states with different eigenvalues for $\mathbf{J}$.

Although spin has a discrete spectrum, it is an external parameter just like energy and momentum. That is, it depends partly on the external reference system with respect to which we orient the system. Now in a physical sense orientation can only be given by a field, which in every spatial point has a certain direction (the z-direction, say). With respect to this direction, we can specify that the l_z component assumes a certain value, whereas the fact that l_z does not commute with l_x and l_y implies that in this case the state of the system (now an eigenstate of l_z) cannot be an eigenstate of l_x or l_y. The system cannot "have" spin components perpendicular to the field.

For some systems (like electrons) the spin is associated with a magnetic moment, such that a magnetic field can orient the spin of the system. For other systems (such as light quanta), the direction of propagation is the determining factor. Light quanta can be transversely polarized in two mutually orthogonal directions, corresponding with $m = \pm 1$ (the third eigenvalue, $m = 0$, can only be realized in longitudinal, "virtual" photons, as a consequence of the typical structure of electromagnetic interaction).

As observed, with every possible direction in space there corresponds a spin-component operator l_z, which does not commute with the other operators, l_x and l_y. Thus if a system is oriented with respect to a certain direction, it cannot at the same time be oriented with respect to another direction. Consider a beam of electrons moving along the x-direction in the presence of a magnetic field pointing in the z-direction. This beam will split up into two parts, because the spin operator l_z has two eigenvalues, $\pm \frac{1}{2}$. If we let one beam pass through another magnet parallel to the former, we will find no further splitting, which shows that after passing the first magnet all electrons in one beam are in an eigenstate of l_z, and remain so before and during their passage through the second magnet. But if the second magnet has its field in the y-direction, the beam will split again in two parts. If we now analyze one of them with a third magnet, again in the z-direction, we will again find a beam splitting, showing that the orientation with respect to the y-axis (in the second magnet) destroyed the earlier orientation with respect to the z-axis.

I think this is overlooked by Ballentine in his thought experiment in which he tries to show the possibility of determining the spin of a particle along two different directions, y and z.[25] His experiment pro-

25. Ballentine; see also Frisch (in: Bastin, 20) on a similar thought experiment; see also Jammer E 235, 302ff, 309.

ceeds as follows. Suppose two particles with $J = \frac{1}{2}$ emerge from a decaying particle of spin zero. This means (because of angular momentum conservation) that the two particles must have opposite spins. If the decaying particle was at rest, the two emerging particles will also have opposite momenta. Now suppose we measure the spin of one with a magnet in the y-direction, and that of the other with a magnet in the z-direction. Then, according to Ballentine we have determined both the y- and z-components of the spin of each particle.

However, I think Ballentine is wrong when he assumes that there is no magnetic field at the place where the original particle decays. Each emerging particle experiences the field of the other. Therefore each particle is already oriented before it arrives at the measuring magnet. And with this magnet, not only the original orientation is destroyed, as we saw above, but also the connection with the other particle.

Ballentine considers the two particles as a single system, which means that their combined state must be described by a single vector in a Hilbert space. But the fact that their spin is oriented relative to each other as described above implies that this Hilbert space for the two-particle system can be separated into two one-particle Hilbert spaces, which are mutually orthogonal, such that a measurement on one system has no bearing on the other. Of course, if the original orientation immediately after the decay can be fixed, there is a statistical correlation between the measurements of I_z for both systems, if the measuring magnets are both oriented in the z-direction.

Ballentine's experiment is proposed as an alternative to the famous Einstein-Podolsky-Rosen paradox,[26] in which the momentum of one system is determined by the momentum of another, with which the former has interacted shortly before the measurement. In this case one makes use of the conservation of linear momentum. But then again, account is not taken of the destruction of the previous state of the measured system due to the measurement of its momentum. This means that nothing can be inferred about the state of the unmeasured system, whose state is orthogonal to that of the measured system.[27]

The objection may arise here that in Compton's experiment it is

26. Einstein, Podolsky, Rosen; see also Bohr A 231ff; Margenau A 261ff; Feyersbend C; Jammer E; Hooker.

27. According to the Heisenberg relations, the states of the two particles after the collision cannot be pure momentum (or energy) states, because of the finite interval (or duration, respectively), in which the interaction took place.

shown that the measurement of the momentum of one colliding particle gives the value of the other one because of the law of conservation of momentum, which therefore is also valid for microsystems. But in this case the value of the momentum is much larger than its statistical spread, so that the latter can be neglected. For low-energy experiments as discussed by Einstein, Podolsky and Rosen, the spread in momentum becomes proportionally more significant. (cp. Sec. 7.6).

For a clear understanding of these paradoxes, one has to keep in mind that the Hilbert space formalism is only applicable to isolated systems. The "paradox" results due to the jump from one isolated system to another. First one considers the two colliding or emerging particles as one system. Then one considers each particle as an isolated system. If we consider the two particles separately, we must then consider them either as interacting with each other (which means they are not isolated), or as having a state prepared by the preceding interaction. But, in the latter case, this state need not be an eigenstate with respect to the measuring instrument. The "projection postulate" (according to which the state of the system changes in a measurement into an eigenstate determined by the measuring instrument) applies to individual systems. Therefore it must not be applied to the state of the two particles together, if the measurement is only done on one of them. It is only the state of the latter particle which is subjected to the projection postulate.

9.7 Typical symmetry
The unitary operators describing the symmetry of a system always form a group. The group of external temporal or spatial translations and the Lorentz or Galilean groups of kinematic motion are continuous. On the other hand, those describing rotational symmetry are discrete, and this is also often the case for internal symmetries.

If we had full knowledge of the internal structure of a system, we would not need the symmetry properties, because they are inherent in the structure. But more often than not, the symmetry relations are better known than the detailed structure, and therefore form a great help in designing and interpreting experiments. This is also the case where the system is in principle known, but mathematically difficult to treat, as in large molecules or solids.

Classically we speak of an internal symmetry whenever it can be determined a priori that for two or more cases a certain probability must be the same, regardless of its precise value. For example, Laplace's "equally favourable cases" are determined by symmetry. But only if all possible cases are equally favourable (as in dice throwing) is it sufficient to find the probability value itself. This is not usually the case in physically qualified systems.

The idea that symmetry leads to equal probability does not rely on a complete "lack of knowledge" or "lack of sufficient reason" as was sometimes assumed in order to reconcile statistical methods with a deterministic philosophy. This idea can even lead to absurdities if taken literally. Take, for example, the case of throwing two dice simultaneously. Beginning with the argument of a "complete lack of knowledge" one should argue that all possible results, 2, 3, . . ., 12, are equally probable, which is patently false.

Symmetry is relational with respect to the environment of the subject concerned. That is, it anticipates possible interaction. It is not an epistemological lack of knowledge, which we confront, but an ontological indifference on the part of one subject (the environment) which interacts with another subject, whose symmetry is being considered.

Now we have seen that in quantum physics the probability of finding a certain state depends jointly on the initial state, and the operator describing the interaction, namely, the reference system. But when we speak of symmetry, we abstract from the initial state. We wish to discuss the symmetry of the *law* for the system, with respect to some reference system. Therefore, contrary to what we stated in the preceding paragraph, we cannot immediately discuss the probability, but instead we shall consider the eigenstates and eigenvalues of the operator in relation to the reference system.

A hermitean operator determines the spectrum of possible values which can be assumed by a certain physical magnitude. Thus we can characterize an eigenstate of the operator by its corresponding eigenvalue, which is the value of the physical magnitude for the system in that state. For example, in a measurement we do not directly determine the eigenstates, but the corresponding eigenvalues, and this is also the case in a state selector. We can manipulate the eigenvalues, but not the eigenstates. This implies that if two eigenstates have the same eigenvalue, we cannot distinguish between the two eigenstates, neither in a selector, nor in a measurement.

The occurrence of two different eigenstates having the same eigenvalue is sometimes accidental. However, this "accidental" degeneracy does not interest us. More important is the degeneracy due to some kind of symmetry. For external symmetries we have already seen that, e.g., the temporal homogeneity implies that an infinity of eigenstates (differing by a phase factor) have the same energy.

Just as in the case of external symmetry (Secs. 9.5 and 9.6) the internal symmetry of a system can be described by a set of unitary

187

operators. If we have just two mutually orthogonal eigenvectors with the same eigenvalue for some hermitean operator, A, then there is a unitary operator U which transforms one eigenstate into the other, and a second unitary operator which performs the reverse transformation and is referred to as the inverse operator U^+. Together with the identity operator, $I = UU^+$, they form a group. If A has more than two orthogonal eigenvectors which are degenerate, then the corresponding group of unitary operators has more than two members. These unitary operators commute with the hermitean operator A, which is therefore left invariant under the symmetry operations. This shows again the extreme importance of group theory for the analysis of typical structures. Although group theory cannot give the eigenvalues of A, it can tell which of them are degenerate.

Now we can look for operators which commute with A (and therefore have the same set of eigenvectors), and which "break" the symmetry leading to degenerate eigenvalues for A. Again group theory can help in solving this problem. It is an important hypothesis that a "complete set of compatible magnitudes" exists for any typical structure. That is, there exists a complete set of mutually commuting hermitean operators, such that each eigenstate can be uniquely characterized by a set of eigenvalues or "quantum numbers".[28] Group theory can show which eigenvalues of some operator of the complete set are degenerate. It is this set of operators which ultimately determines both the full set of possible states, and their relative weights (jointly with the initial state of the system).

Such a complete set of operators is not unique, however. There are always several mutually incompatible sets of compatible magnitudes. Thus, any component of the spin can belong to such a set, but different spin components are mutually incompatible, and cannot simultaneously belong to the same set.

There is no criterion known for the completeness of such a set. The possibility always exists of finding a new kind of symmetry applicable to the system. For example, for a long time it was thought that a complete set of magnitudes for a free electron consists of momentum (or energy as an alternative), mass, spin, and one of the spin components. Later it was realized that the electron could exist in two charge-eigenstates, one negative for the "ordinary" electron, and one positive for the positron. Thus charge must be added to the set. Recently, yet another attribute, the lepton number L, has emerged, and is compatible with all the members of the set.[29]

28. Messiah 294.
29. Kaempffer 2.

This leads us to another extremely important point, already alluded to in the preceding section. This is, namely, the application of the superposition principle, which is the foundation of the Hilbert space concept. If two eigenstates of some operator (such as the Hamiltonian in connection with the energy) can be superposed, then the system is said to be in the state objectified by the linear superposition of those two eigenstates. In particular, if two eigenstates, for some operator, are degenerate, then any linear superposition of the two states is also an eigenstate of that operator. One consequence of the complete set of compatible operators is that no two eigenstates can be degenerate with respect to all operators of the set.

However, not all eigenstates can be superposed. For instance, of the properties mentioned above for a free electron, only the eigenstates of the momentum and of the components of the spin can be superposed. The mass, charge, total spin, and the lepton number are subject to so-called "superselection rules", stipulating that their eigenstates cannot be superposed. So though we consider the electron and the positron as different states of the same system, these states cannot be superposed. There are no states with partly positive and partly negative charge. This distinction between superposable and non-superposable properties allows us to distinguish between two basically different kinds of typical structures (cf. Sec. 10.5).[30]

In conclusion, we find the somewhat surprising result that in quantum physics the symmetry of the system does not immediately lead to a prediction concerning the probability of a state, but rather to one concerning a possible value. If, according to symmetry, different eigenstates have the same eigenvalue, we cannot distinguish these states by a measurement that discriminates on the basis of the eigenvalues. If the initial state is fixed (e.g., by a state selector), then the probabilities associated with the different eigenstates possessing the same eigenvalues may be quite different.

However, if, for the sake of argument, we assume the initial state to be arbitrary, then all eigen*states* have the same probability, and thus the relative weights of the eigen*values* are proportional to their degree of degeneracy. And this is once again the same as in classical physics, which usually assumes a completely random initial state.

---

30. In more complicated systems like atoms there are also "selection rules" which sometimes refer to the same kind of operators (total angular momentum, e.g.). However, the transitions which are "forbidden" by these rules are often simply less probable than other transitions because the symmetry underlying these rules is "broken" by some perturbing interaction; see any textbook on quantum physics.

9.8 The temporal evolution of an isolated system

In Sec. 9.5 we discussed the conservation laws of energy and momentum with respect to an isolated system, and we found that its external motion is just the plane wave motion with which we dealt in Chapter 7. This is the case e.g. for electrons. For complicated systems, like atoms, the energy operator, the Hamiltonian H, not only has relevance to external motion (kinetic energy), but also to internal motion. Then it is comprised of potential energy as well as kinetic energy due to the relative motions of the electrons and the nucleus in the atom.

Now it is sometimes called the "central dynamical postulate" that for any isolated system the temporal evolution of its initial state is determined by the Hamiltonian. In particular, if the initial state happens to be an eigenstate of the Hamiltonian (i.e., if it is prepared by a pure-state energy selector), the state will remain the same, if it is a stationary state, with a definite eigenvalue, the energy of the system. Strictly speaking, only the ground state of an atom is stationary, and any other state will sooner or later decay. Implicit in such a decay process, however, is the occurrence of a reaction to the environment, which means the system is not strictly isolated.

Nevertheless, the Hamiltonian can still tell us something about the probability of decay, i.e., the mean decay time. An unstable state can always be conceived as the superposition of several stationary states, and the Hamiltonian determines the probability that the system will be found in a state consisting of the ground state of the original system plus a free photon. The only requirement is that the energy of the system in its ground state plus the energy of the emitted photon be equal to the energy of the initial unstable state. This is only an approximation, however, and a more satisfactory theory is found in quantum electrodynamics, which starts from the interaction between the system and an electromagnetic field. If the initial state is not an eigenstate of the Hamiltonian, then its change will be determined by its relation to the electromagnetic field.

If the initial state is a mixture, it can be shown that it becomes more and more mixed. In fact, for mixed states a magnitude can be defined having the same properties as the classical entropy. For a pure state, this magnitude is zero whereas for a mixed state it is positive. For an isolated system it increases during its temporal evolution, which is therefore subjected to the physical time order of irreversibility.

We have already mentioned several distinctions between pure and mixed states. Another one is found in interference, the most characteristic kinematic property of waves. Interference is only possible if (and as long as) the interfering waves are "coherent", i.e., if their phase relations are not at random, but determined by some previous interaction, namely, the preparation of the wave packet. Thus one speaks of the "coherence length" of a wave packet. Now a pure state has an infinite coherence, and a mixed state has a finite one. The possibility of interference is therefore limited for a mixed state (and of course, every actual state of any concrete system is more or less mixed). Thus any distinction between pure and mixed states has to do with irreversibility as the physical time order, and is related to some interaction. It turns out that the pure state is an idealized boundary case, just as a state of rest is a boundary case of motion.

We have not yet commented on the fact that we describe the isolated system with a complex Hilbert space. Most authors on the foundations of quantum physics take this for granted. Tolman calls the "phase factor randomness" (cf. Sec. 9.4) the basic postulate of quantum mechanics,[31] but Jauch states that as yet no one has shown that a real Hilbert space fails to do the job.[32] I hold that complex numbers are most suited to a description of interference phenomena, which is one of the characteristic differences between classical and modern (opened-up) kinematics.

It is important to realize that the Hilbert space concept is strictly confined to isolated systems. This implies the possibility of describing the internal structure of systems like nuclei, atoms, molecules and solids, and also simple cases of collisions (which are describable as isolated events). The theory allows one to incorporate internal interactions. The problem of the actualization of possibilities is then circumvented, because one restricts oneself to the calculation of stationary states and mean values of properties. As to its external relations, the theory can do little more than describe the particle's motion. With respect to external interactions, it can only give probabilities, in an anticipatory way. The kinematics of quantum physics is one which is opened up.

9.9 *Actualization*
Thus we find that the Hilbert space formalism has a very wide scope since it can solve four of our problems mentioned in Sec. 9.1. But it has its limitations too. It cannot solve the fifth problem, concerning the actualization of one of the possible states.[33] This is often called the "measurement problem", although it pertains to any external interaction. This is not always recognized, because in many cases of

---

31. Tolman 349ff.
32. Jauch 121; see also Weizsäcker B 252.
33. Feyerabend C.

external interaction, it is possible to solve the problem by including the two interacting subjects into one single system, in which case the Hilbert space formalism is applicable. This method is very successful in describing the interaction of, e.g., electrons and nuclei, and somewhat less successful in the description of the interaction between an atom and an electromagnetic field. The latter does not lend itself for treatment as an isolated system.

In order to see the relevance of interaction to the understanding of our problem, let us consider the following example. It is a well-known statement that an atom has discrete energy levels, each corresponding with an eigenstate of the Hamiltonian operator. For the sake of argument, let us only consider two such states, the ground state and the first excited state. The Hilbert space formalism says that the atom may also be in a mixed state, described by a Hilbert space vector which is a linear combination of the two eigenstates. To this mixed state there corresponds no fixed energy value. How can we interpret this?

As long as we consider the atom as isolated, we are at a loss, because every isolated system has a fixed energy. However, no system is really isolated. We can study this problem with the help of quantum electrodynamics, observing that each atom interacts with an electromagnetic field, such that the mixed state can be considered as the ground state together with a virtual photon – in which case the energy of the atom is not fixed, but neither is the atom isolated.

There is also a probabilistic interpretation. In fact, we shall never have a single atom, but rather can consider a dilute gas. The interaction between the atoms may be so small, that for the calculations we may assume that the atoms are isolated. At the same time, the interaction may be large enough (e.g., via collisions: the temperature of the gas must be well above zero) in order to have atoms both in the ground state and in the first excited state, in a ratio determined by the Boltzmann factor (cf. Sec. 8.5), which therefore gives the probability that a certain atom is in one of these states. This ratio is immediately related to the mixed-state vector, for which we sought an interpretation. Thus by introducing a weak interaction, small enough not to disturb the calculational results, we can give a very natural interpretation of these mixed states.

Also in these cases, however, no actualization is described, and the success of the theory lies in the predictability of observational probabilities, stationary states, etc.[34] It has been suggested that the main reason why the Hilbert space formalism fails is based on the super-

34. Popper C 25, 26.

192

position principle, by which the linear combination of two possible states is again a possible state. Hence, if a state is a linear combination of two eigenstates for some operator, the system can be said to have simultaneously two different values for the magnitude represented by that operator. This principle is only applicable to *potential* states, not to actual ones. Whereas the possible energies of a system can be superposed, as can be verified in interference experiments, actual energies cannot.

We have already met this state of affairs in Sec. 9.7 when we distinguished between superposable and non-superposable properties of isolated systems. Only for the latter can we say that the system "possesses" the property. Thus an electron "has" a negative charge, a positron "has" a positive charge, but neither "possesses" a certain energy value. Only if the electron has actualized one of its possibilities in some external interaction, may it be said to "have" a certain energy. This is not only the case in a measurement or observation, but also e.g. in a collision between two electrons, if after the collision takes place, there is a correlation between their states. Evidently a theory of external interaction must do away with linearity and superposability. This was recognized some time ago by several physicists, among them Heisenberg, De Broglie, Vigier, and Bohm.[35] However, the mathematical difficulties involved in a non-linear theory are enormous, and the theory is progressing only very slowly, if at all.

So time and again we find that the conceptual difficulties involved in quantum theory can be understood if we keep in mind that the theory only applies to *isolated* systems and *potential* interactions. This poses the question as to what extent the concept of an isolated system is fruitful. Strictly speaking, isolated systems do not exist, and the concept itself is even objectionable, since we characterize physical systems by their interaction as the basic physical subject-subject relation (cf. Chapter 5). We stress again that it is a subject-subject relation, an interaction between two physically qualified systems, which is at stake. It is not a psychically qualified subject-object relation between the observer and the observed system, as was believed by Bohr and Heisenberg, e.g.[36] I do not say that the latter relation is of no interest. However, if one wishes to study the latter, one must first of all be aware of the former subject-subject relation and its implications, precisely because any observer (a psychic subject) is a physical subject as well.

The answer to our question must be found in the study of the individuality structures which are physically qualified. As we shall see

35. Bohm, in: Bastin 102ff.
36. See, e.g, Heisenberg A 2; Heitler 191.

presently, both the driving motive and the success of quantum theory must be sought in this field of research. But it also gives us a clue to the understanding of our problem.

First, it is just the (limited) *individuality* of physical systems which makes it both necessary and fruitful to consider them in isolation, in an intermediary state between two interactions. Thus the weakness of quantum theory is its strength all the same. Secondly, the distinction of potentiality and actualization will be related to the distinction between the "thing-like" and the "event-like" character of physically qualified individual subjects.

10. Structures of Individuality

10.1 *The main subject matter of quantum physics*
Quantum physics has emerged from atomic physics and, at times, the two are still identified. Quantum formalism (as discussed in Chapter 9) however, has been applied successfully to molecular physics, solid state physics, astrophysics, nuclear physics and (less successfully) to sub-nuclear or high-energy physics. The lack of success in the latter case generally is not ascribed to shortcomings in quantum physics (although some speculations have been made to this effect), but to the largely unknown nature of the weak and strong nuclear interactions.

From a systematic point of view, quantum physics can be divided into three parts: the wave theory of motion (Chapter 7), the theory of probability (Chapter 9), and the investigation of the typical structure of matter. The quantum mechanical wave theory is basically not different from classical wave theory, except for the – indeed very important – recognition of the relation between energy and frequency, and the recognition of the general, modal character of the wave theory. The quantum theory of probability differs from the classical one because of the possibility of interference, and because of the new insight into the law-subject relation for stochastic processes.

These differences between classical and quantum physics are well-known, and widely discussed. But the third difference, the fact that classical mechanics as a modal theory of motion is unable to account for typical structures has usually been overlooked. Neither can the modal theory of wave packets or the general theory of probability account for typical structures. To describe the typical structures of atoms and molecules, which came to the physicist's attention not long before 1900, a completely different theory is required. Such, in our opinion, is the nature of quantum theory.

Thus, in our view, quantum physics did not begin with the discovery of Planck's constant in 1900, nor even with Einstein's conjecture regarding the quantum nature of light in 1905. Rather, quantum physics began with the study of the atom by the spectroscopists, by Rutherford, Bohr, Sommerfeld, and others. Bohr, for example, considered the stability of the atom to be the basic problem. By pursuing this problem as the central one, in 1913 and later, Bohr made his most remarkable contributions to the development of quantum physics. He made it clear that atomic stability cannot be accounted for by electrodynamics in conjunction with classical mechanics.

In this respect Bohr diverted sharply from the views of most of his contemporaries, including Planck and Einstein. Planck continuously sought a reconciliation of the new experimental and theoretical results with classical physics, whereas Einstein, in a more revolutionary mood, was searching for a new unified theory embracing both electrodynamics and mechanics which would account for the new phenomena. In contrast, Bohr's approach alone turned out to be fruitful, and so it is that he should be seen as the principal originator of the new theory.[1]

This is not to imply that we can dispense with the contributions of Planck and Einstein, for the study of the typical structures depends on an insight into the opened-up structure of the modal aspects involved. We have seen in Chapters 7 and 9 that this unfolding results in the wave theory and in the theory of probability, both of which serve as a necessary basis for the theory of typical structures.

We shall freely speak about electrons, atoms, nuclei, etc., assuming their individual reality. We view them as no less real than, e.g., rocks.[2] In objection, nominalist physicists and philosophers would argue that, since we cannot "see" such species, we should consider them (instrumentally) as theoretical constructs,[3] whose existence is not directly given, and therefore is not equivalent to the existence of directly visible or tactile things.

It may be noted, however, that such considerations (to my knowledge) are never directed towards the existence of stars and other celestial subjects. Yet we see the stars merely as point-like sources of light even with the most powerful telescopes. In much the same way we are also able to "see" electrons, e.g., as light spots on a T.V. screen. In fact, everything we know about stars (their mass, density, volume, chemical constitution, life time, etc.) has been discovered in precisely the same way as our knowledge of electrons, atoms, etc. In both cases, our knowledge is derived from instrumental measurements with the help of experimentally confirmed theories. Thus, if electrons are theoretical constructs, so are the stars.

It could, of course, be argued that celestial subjects can be in-

1. Meyer-Abich 14, 15; Weisskopf 42, 294f; Heilbron, Kuhn. For a different view, see Popper C.
2. Weisskopf 35; Cantore 207ff; and Grandy's anthology.
3. Margenau A, Ch. 4; Margenau, Park; Toulmin 120ff; Bridgman 59.

vestigated in ways other than with telescopes, e.g., with space probes. But then, with respect to the sub-microscopic subjects we must remember that in the 19th century there were physicists and philosophers (notably Mach and Ostwald)[4] who contested the individual existence of atoms and molecules. It was not until Einstein and Smoluchowski gave a theoretical quantitative explanation of Brownian motion, and Perrin used it to determine experimentally Avogadro's number (the number of molecules in a mole), that most physicists and philosophers became convinced of the reality of atoms and molecules.

Currently, it is already possible to obtain photographs of large molecules and even of separate atoms with a field-ion microscope. Thus the boundary has shifted. Just as we now know that the back of the moon exists, so, too, the existence of atoms and molecules is now unquestionable. Today, only electrons and other sub-atomic particles are considered, by some, to be "theoretical constructs".

In Sec. 8.1 we already stressed that the individuality of physically qualified things and events cannot be proved as the result of a theoretical investigation. In this sense defenders of the "theoretical construct" argument are right: theories create concepts, not real things and events. Therefore we reject nominalism as well as any kind of "realism" which attempts (mostly on speculative metaphysical grounds) to provide a philosophical basis for the reality of an "outside world".

We are forced to borrow the idea of individually existing subjects from pre-theoretic, daily experience. Atoms, electrons, etc., are not encountered in daily experience and, therefore, their individuality must be inferred by extrapolation. Once the existence of individual atoms or smaller systems has been established (on the basis of a variety of evidence), they should be considered as things which simply are smaller than the smallest things we can see with ordinary means. Similarly, we consider stars to be individual subjects on a par with the sun, the only difference being that our distance from the stars is much greater than that from the sun. In other words, the fact that our awareness of atoms and electrons, stars and galaxies, or of even less firmly established systems like quarks and quasars, has emerged from theoretically aided observations, does not imply that their existence is of a theoretical origin, but is solely a result of their very minute dimensions or their exceedingly vast distances.

The view that electrons etc. are theoretical constructs, if taken consistently, inevitably leads to agnosticism which is an irrefutable world view. Such a view can only be countered by the pre-scientific

4. Mach 588-590.

197

conviction that the individual existence of anything real outside of ourselves is guaranteed by the faithfulness of the Creator to his creation. Therefore, we call the typical structures of creation, "structures of individuality", which, for physically qualified systems, are characterized by their ability to interact with each other.[5]

In conclusion, we state that quantum physics is concerned primarily with the theoretical investigation of physically qualified typical structures. In our introductory chapter we mentioned that one aim of science is the reconstruction of structural laws from modal and typical relations (page 13). The order of such an investigation requires initial abstraction from individuality and typicality in order to discover the modal aspects and their mutual retrocipations and anticipations. One then returns to the typical side of reality, in an effort to analyze the typical structures encountered there.

As it turns out, the modal aspects by themselves are far too general for this purpose, for no typical law can be derived from modal laws alone. However, besides the purely modal gravitational interaction, there are a number of interactions which, though they cannot be considered universal in a modal sense, have sufficient generality that further specification leads to an enormous number of different structures (Sec. 10.2).

After such specification has been made, the reconstruction (the theoretical synthesis) of physically qualified typical structures in modal terms requires that these structures, as we understand them, still display the modal diversity of the first four modal aspects as we discovered them in the preceding chapters of this book. All physical structures are qualified by the physical modal aspect of interaction. In addition we shall distinguish three types of typical structures – particles, enkaptic wholes, and quasi-particles. Each of these is founded on a different pre-physical modal aspect, of which there are three – numerical, spatial and kinematic. This distinction, with some of its consequences, will be the subject matter of Secs. 10.3 to 10.8. It allows us to account for the typical diversity within the "kingdom" of physical systems.

10.2 *Typical interactions*
Until 1930 it was thought that there were only two fundamental kinds of interaction, gravitational and electromagnetic, and it was hoped that they could be unified in a single theory.[6] In fact, for some time

5. Cantore 210: "As far as working physicists are concerned, atomic objects are real because they can be shown experimentally to occur with definite properties of interaction".
6. Jammer D, Ch. 11, 14; Popper C 8, 9.

general relativity was considered as the basis of a so-called "unified field theory", a modal theory within which, supposedly, every kind of interaction could find its place. But soon after 1930, physicists realized that the interaction between protons and neutrons cannot be merely one of an electromagnetic character, since the strong nuclear interaction between nucleons is even independent of their charge. In addition to this strong nuclear interaction, which is clearly irreducible to electromagnetic interaction, one can also distinguish a weak nuclear interaction, although there are some reports which link this interaction with electromagnetic interaction.

The typification of a modal subject-subject relation is not found in the mathematical modal aspects. Whatever the typical structure of two subjects, their numerical, spatial or kinematical relations, insofar as they are not determined by the physical modal aspect, are of a general, modal character. In the physical modal aspect we have established that only gravitational interaction is a general, modal interaction (Sec. 4.7). In thermodynamics, the internal energy of a subject is the sum of a number of different types of energy, for example, mechanical potential energy, magnetic energy, electrical energy, chemical energy, etc. They give rise to typical forces and typical currents. It is assumed that all these types of energy can be reduced to one of the "fundamental" interactions – gravitational, electromagnetic, strong or weak nuclear interactions.

This typification of physical interaction is a prerequisite for the possible existence of typical structures. These are structures in which typical interactions are bound together in a typical way, although typical interactions also function in mixtures and unbounded systems in which different typical systems interact without forming a new structural whole. The typification is also conditioned by the individuality structures of the interacting systems. A specific kind of interaction like the electromagnetic one is determined not only by the modal relations of the interacting subjects, such as their relative motion and distance, but also by the typical nature of the subjects involved.

The typical interaction has both a law side and a subject side. In the case of electromagnetic interaction, the law side is expressed by Maxwell's equations whereas the subject side is expressed by the typical charge, magnetic moment, etc., of the interacting subjects. For nuclear forces, we appear to have greater insight into the subject side of such interactions than we do of the law side. It appears that this is largely due to the difficulty of disentangling the interactions of many particles into pairwise subject-subject relations. As we have seen, it is precisely this kind of relation which provides a powerful element in the analysis of the modal and typical structures of reality.

Each of the typical interactions has its own kind of symmetry.[7] Electromagnetic interaction is subjected to the so-called gauge-invariance, which leads to the law of conservation of charge. Recently, similar invariance laws of "baryon number" and "lepton number" are developed. In addition to these and the well-known laws of conservation of energy, linear and angular momentum, conservation laws have been discovered for "parity", "isospin", "strangeness", "hypercharge", etc., which are usually shared by all fundamental interactions, though they are occasionally "violated" by one of them.

The "elementary particles" are usually classified according to these properties. Thus we distinguish baryons (interacting via all fundamental interactions, and having half integral spin), mesons (all interactions, integer spin), leptons (no strong nuclear interaction, half integer spin), and some solitaries such as the photon and the graviton (no strong interaction, integer spin). Although the four fundamental and irreducible interactions can be ordered according to their strength, there is also some indication that they can be ordered according to their symmetry. Thus, the strong nuclear interaction appears to have the greatest internal symmetry, whereas gravitational interaction has the least.[8]

Of the four fundamental interactions, it seems that electromagnetic interaction is most important. The nuclear forces are capable of acting over only very small distances, which means that they are relevant only for nuclear or sub-nuclear structures. Whether or not gravitational interaction gives rise to typical structures is not clear. There is some indication that certain stellar structures owe their typicality to gravitation, but these theories are even more obscure than the sub-nuclear ones.

Because philosophical reflection on the foundations of a special science is necessarily retrospective, it should therefore refrain from speculation. Hence we shall base our following considerations primarily upon well-established theories concerning the structures of nuclei, atoms, molecules, and solids as well on some general properties of sub-nuclear particles, in addition to the interactions between

---

7. See, e.g., Ovchinnikov, in: Kuznetsov, 89-140.

8. Ovchinnikov, in: Kuznetsov, 127. These typical differences between the fundamental interactions are relevant for all three retrocipations of interaction and not only for the strength of the interaction. Thus the electromagnetic field has a vector character, the strong nuclear force corresponds to a scalar field, and gravitation is a tensor field. The typical currents have corresponding typical differences: the electromagnetic current is a transverse wave and the strong nuclear current is a longitudinal one.

these systems. We shall not enter into speculations on possible future developments.

Later, we shall distinguish between "particles" and "enkaptic wholes", like nuclei, atoms, molecules, and solids. Traditionally, the fields of inquiry dealing with these structures are divided according to the "quantum ladder",[9] with very small systems corresponding to high energies per individual particle. However useful such classification systems may be, from a systematic viewpoint they are too obvious. Within the framework of our philosophical approach, we shall suggest an alternative classification of typical structures, not with the intention of replacing the more traditional ones, but to arrive at a deeper understanding of the principles of the theory. This alternative classification is orthogonal to the preceding ones in that it is related to the anticipation-retrocipation distinction.

10.3 *The thing-like and the event-like*

No typical structure can be grasped in a single modal aspect since each structure of individuality always embraces all modal aspects in a typical way. Thus any individual has a numerical character in its typical *unity*, which is the reason why we are able to count individuals of a kind. Each individual system is characterized by *properties*, which is an expression of its spatial character, since we can only speak of a property if it points to a spectrum of possibilities. Each individual is also subjected to *change* (kinematic) and is able to interact (physical) with other systems. Thus each individual has *subject functions* in several modal aspects, and it has *object functions* in others. Each typical structure has a qualifying modal aspect, which may be a subject function, as is the case for all physically qualified structures, or an object function, as is the case, for example, for an art object.

In physics, we are concerned only with subject functions of physically qualified structures. This does not mean, of course, that we may neglect all object functions. In particular, the observational and cognitive aspects of these structures which enable them to become objects of study experimentally and theoretically are always important. The common property of all physical systems – namely, that they are all physically qualified structures, and thus have subject functions in the first four modal aspects – delimits the physical sciences from other disciplines such as biology or psychology.

Precisely because all physical structures have subject functions in the same modal aspects, and object functions in all others, their modal behaviour is very similar, and can in itself not be used for

9. Weisskopf 41-51.

further classification. Thus, for example, the wave properties discussed in Chapter 7 are possessed equally well by electrons, photons, and alpha-particles. Unfortunately, similarities of this sort have tended to eradicate the fundamental distinction among structures of this kind.

We shall propose a new classification scheme which has a twofold character. First we shall show that each typical *thing-like* structure, besides the qualifying physical aspect, possesses another *foundational* modal aspect. This will enable us to distinguish three thing-like types. We shall distinguish between physically qualified structures which possess a numerical foundational modal aspect (which are numerically based); those which possess a spatial foundational aspect (spatially based); and those which possess a kinematic foundational aspect (kinematically based).

These distinctions are of a *retrocipatory* character, and are therefore related to the retrocipatory analogies of physical interaction, viz. energy (numerical), force (spatial), and current (kinematic) (cf. Chapter 5). Thus we can understand the thing-like structures as individualizations of energy, force, and current, respectively. It allows us to give an objective representation of the physically qualified structures of individuality (Sec. 10.4).

Next we shall see that this distinction is insufficient for a complete understanding of the functioning of these structures. This is due to the fact that this thing-like classification is of only a retrocipatory nature, whereas we can fully understand the typical structures only in their opened-up, anticipatory functioning. This has an *event-like* structure.

Whenever we speak of a "thing" we explicitly or implicitly refer to its permanence on the law side, and its individual duration as a unity on the subject side. The individual existence of a thing is rooted in its *permanence*. For spatially extended systems, we must also assume that the system has a certain *coherence*, whereas if internal (periodic) motions are possible, they must be such that the system is none the less *stable*. Therefore, permanence, coherence, and stability are retrocipatory aspects of a thing-like structure.

But the thing-like description fails as soon as we are confronted with unstable, incoherent systems with a limited life time. This can best be illustrated by the concept of energy. For a stable system, the energy has a real value. It is the eigenvalue of the Hamiltonian operator. But for a system with a finite life time, the energy turns out to be a complex number, and the imaginary part describes the probability that the system will decay after a certain time interval. This probability is usually expressed as a mean decay time which is related to the relaxation time (cf. Sec. 7.1). As a result, this decay can only be understood in the anticipatory direction and it clearly describes an event. A physically qualified event can thus be described as a typical transformation of energy. Similarly, the quasi-particles can be understood in thing-like terms as long as they move from one system to another, but, if we consider their emission or their absorption, they can only be described in an anticipatory (event-like) manner.

The thing-like character of molecules in their ground state depends on the fact that events cannot easily occur and this can be explained by the discrete energy spectrum. Since the mean kinetic energy of molecules at low temperatures is much smaller than the excitation energy for the lowest excited state, it is possible to consider collisions between molecules in a gas as being elastic and not changing the internal state of the molecules involved. This accounts for the remarkable success of the 19th-century kinetic theory of gases (Sec. 8.5). However, as soon as the mean kinetic energy of the molecules becomes comparable to the excitation energy, or especially the ionization energy of the molecules, we can no longer consider them to be stable thing-like structures, even if they have a potential energy minimum. This is the case, for example, in iodine at ca. 1000 K. In stars like the sun, the mean kinetic energy is so high that no atoms or molecules exist for long in their respective ground state.

We must warn the reader that we intend nothing metaphysical in the distinction of thing-like and event-like. We imply no "Ding an sich", nothing "eternally existing and unchangeable", nor even a sharp separation of classes of things from classes of events. Electrons, atoms, photons, etc., have both thing-like and event-like features, and in a given context we may be interested in one or the other.

The thing-like may also be referred to as "being" and the event-like as "becoming", but here too, care must be exercised with the use of these terms, for they are bound to cause confusion since no key terms have the same meaning in different philosophies. That the philosophical problem touched upon here is very old indeed will be clear if we recall Parmenides' philosophy of being, and Heraclites' philosophy of becoming.[10] The thing-like and the event-like should be seen as complementary components, since, in isolation, neither can give us a full insight into the typical structures studied in physics. In a very essential sense we ought to consider the thing-like and the event-like as *temporal* being and *temporal* becoming.

10.4 *The individualization of energy, force, and current*
As we have seen in Chapter 5, energy, force and current are retrocipatory analogies of physical interaction and also have a typical nature

10. Russell C, Ch. 4, 5; Reichenbach C, Ch. 26; Weisskopf 198-229.

with respect to typical interactions. We have already discussed this matter with a view to the modal side of reality. On the typical side of reality we find that physically qualified structures must have pre-physical foundational functions. Thus we can distinguish:

- material particles, such as electrons, characterized as typical individualizations of energy or quantity of interaction;
- enkaptic material totality structures, such as atoms, characterized by a typical individualized force or another spatial measure of interaction;
- structures usually called "quasi-particles", such as photons, characterized by a typical individualization of a current or interaction in flux.

These distinctions must not be taken too rigidly. Thus, an enkaptic typical whole may appear as a particle in a new enkaptic totality structure. For instance, in a crystal or molecule, an atom (itself an enkaptically bound structure) is enkaptically bound in the new structure and appears as if it were indivisible, i.e., in the crystal or molecule, the atom can be treated as a "particle". Similarly, in the structure of an atom the electrons and the nucleus can be treated as "point charges" (particles) even though the nucleus itself is an enkaptically bound structure composed of a number of nucleons (protons and neutrons). Again, the nucleons themselves are presumably enkaptically bound structures composed of quarks bound up by gluons in a manner that is not yet clear.

It is becoming apparent that the "most elementary" particles, if such entities exist at all, have no independent existence, but can exist only in relation to other such particles bound up in an enkaptic typical whole. This seems especially true for the existence of quarks, three of which make up baryons and two of which form mesons. Thus far, some disputed reports notwithstanding, independently existing quarks have not been observed.

This complication does not affect the structural distinction of particles and enkaptic typical wholes. However, we should understand particles apart from any metaphysical notion of substance. Particles are not like the Kantian "Ding an sich". The notion of a particle has meaning only if it is referred to its physically qualified relation to other structures, and if this relation is numerically based, the particle can be understood as a typical individualization of energy. Specifically, the particle has a typical rest mass, in addition to typical charge, spin, etc., which are quantum numbers subjected to superselection rules.

In an enkaptic typical whole, the energy is individualized into discrete energy levels, corresponding to eigenstates of the Hamiltonian operator. These eigenstates can be superposed. We shall see that the superposition and superselection rules provide a sufficient criterion by which to distinguish material particles from enkaptic wholes. The internal structure of an enkaptic whole is described by the stationary states of the Hamiltonian operator, which describes the typical law for the system. The Hamiltonian is the sum of a kinetic energy operator and a potential energy operator. The latter describes the forces between the parts of the system and determines the typicality of the system which distinguishes it from other systems. Thus the structure of an enkaptic typical whole is determined by a typical force or field. The structure of an atom, for instance, is characterized by the central field of force of the nucleus interacting with a number of electrons. Stationary states in a system exist only if the potential energy is time independent, i.e., only if the forces acting on the system are static.

Finally, physically qualified "quasi-particles" are characterized by individualized currents. For example, the light quantum or photon can be understood only as emerging from an atom or molecule or as being absorbed by such systems. Therefore we do not consider the light quantum to be a physically qualified material particle. Both its quantity of energy and its extension are determined entirely by the interacting atoms. The concept of a quasi-particle plays an increasingly important role in quantum electrodynamics, high energy physics, and in solid state physics. In solid state physics alone it is possible to distinguish some ten different sorts of quasi-particles already. Just as in the case of enkaptic wholes, there are circumstances in which quasi-particles may themselves be considered to be particles, as is likely the case with mesons, for example. For other kinds of quasi-particles, however, this seems impossible.

While the external motion of a typical whole such as an atom is similar to that of a particle (a free electron and a free atom can both have any velocity with respect to another moving subject), this is not the case for light quanta (photons) or sound quanta (phonons). In either case, the velocity of the wave is independent of the velocity of the source. The speed of sound waves is a characteristic of the medium in which the wave is propagated. Thus the measured velocity of sound depends on the nature of the propagating medium, as well as on the velocity of the measuring instrument with respect to that medium. Since light propagated in vacuum has no medium of propagation, the speed of light must be independent of the velocity of the reference system (cf. Chapter 4). Thus, in contrast to particles and enkaptic wholes, quasi-particles, as individualizations of currents, have a typical velocity.

The individualization of energy, force, and current implies a simultaneous individualization of the modal temporal relations to which they respectively refer, viz. numerical time difference, spatial relative position, and kinematic relative motion.

The numerical time difference is individualized as an individual numerical *time duration* – the temporal "extension" of the subject in opened up numerical time. This time duration, which anticipates the physical interaction, is continuous, uniform, and irreversible. It is correlated with the typical relative *persistence* or *permanence* of the subject on the law side, and is related to its internal energy. If, according to its typical law, the system is stable, then its mass has a fixed numerical value. If the system is not stable (as is the case in a radioactive nucleus or an excited atom), then its mass does not have a fixed value, and the spread is related to the typical decay time of the system.

The spatial relative position is individualized as an *individual spatial extension*, which is correlated with the typical *coherence* of the system. We cannot speak meaningfully of the individual extension of a thing if it is not coherent. In an enkaptic typical whole, the individualized force determines its coherence. The mutual distances of the enkaptically bound particles are delimited, but not completely determined, and the particles preserve their own individuality.

The kinematic relative motion is individualized as *periodic internal motion*, correlated with the typical internal *stability* of the system. Once again, the relative motions of the enkaptically bound particles are restricted but not completely determined, in this dynamic state of stability.

10.5 *The concept of a particle*

Ever since the Greek atomists Leucippus and Democritus, the concept of a particle as an ultimate material unit has occupied a key position in physical thought.[11] It served to reconcile Parmenides' conviction of being as unchangeable (any change was nothing but appearance) with the experienced fact of a continuously changing world. In our terminology, the unity of being (the unchangeable) was relegated to the spatial modal aspect, and change to the kinematic aspect. Until the end of the 19th century atoms were considered as having two traits: they were unchangeable, indestructible, with fixed extension, as well as capable only of motion, thus undergoing collisions and interaction. Such views leave room for several variations. Thus, in ancient thought, the atoms could have different shapes

---

11. Van Melsen B; Dijksterhuis; Harré; Jammer D 338-345; Jaki, Ch. 4; Weizsäcker A 33-50; Weisskopf 198-229; Heisenberg C, Ch. 4; Reichenbach D, Ch. 11; Kuznetsov et al.

and extension, but were all made of the same stuff and they moved in the void. Descartes rejected the idea of a void and considered three kinds of matter, trying to reduce all kinds of change to vortical motion. He was the first to endow the atoms with mass. Newton and his followers adhered to a force-matter dualism, in which everything was explained by forces acting between point-atoms. During the 19th century, the accent of atomic theory shifted to the kinematic aspect.

Particularly in the 19th century, a particle was thought to be characterized by an invariant mass and by an instantaneous position and momentum. This view was shaken by Einstein's special theory of relativity showing the equivalence of mass and energy and their frame dependence, and establishing that light, too, should have mass (besides energy and momentum, which was already recognized in the late 19th century). Wave mechanics, implying Heisenberg's indeterminacy relations, also undermined the traditional view. Nevertheless, even modern authors still hold the view that a particle must be localizable in space and time.[12]

Thus far we have discussed the "philosophical" concept of a particle.[13] However, since the close of the 18th century, it became clear to chemists that the notion of atoms endowed with specific properties was a potentially fruitful idea. Thus the development of the "scientific" concept of a particle was due to chemistry, and this achievement is not devalued by the eventual discovery that the chemical atoms are not indivisible. When Maxwell and Boltzmann applied the scientific concept of an atom to gas physics, other physicists (notably Mach and Ostwald) continued to attack this application on philosophical grounds. Nevertheless, the scientific attitude prevailed and the reality of atoms, electrons, nucleons, etc. is now generally acknowledged, not because of philosophical arguments, but because physicists are able to determine their typical properties, both theoretically and experimentally.

Both traits of the "philosophical atom" were lost. Restricting our considerations to the electron for simplicity, we note that it was realized, even before the advent of quantum physics, that any attempt to endow this particle with some sort of fixed extension would encounter serious difficulties. Such efforts are now considered to be doomed to failure. This means that in the retrocipatory direction,

12. Margenau A 320ff; Jauch 195ff; Cantore 212: "Probably the most common objection raised against the concession of reality to atomic objects is the one which concerns the reality of local motion at the atomic level".
13. The distinction of a "philosophical" and a "scientific" particle can be found in Van Melsen B.

spatial extension has no direct meaning for particles, since, as a thing, a particle is numerically founded. As such, it has a well-defined mass and electric charge. Only in the anticipatory direction is it meaningful to speak of the extension of a particle, i.e., as a collision cross section, anticipating interaction with other systems. Thus, it is not very surprising to find that the collision cross section may depend on the typical structure of the system with which the electron interacts as well as on its relative velocity with respect to the electron. Similarly, the kinematic trait of the classical atom has disappeared in modern quantum theory.

We conclude that a particle is neither kinematically nor spatially founded. Rather, in our view, it has a numerical foundation and is physically qualified. This view does not imply a return to the notion of the particle as a "quantity of matter". If anything, "matter" is a metaphysical concept, not endowed with any physical meaning, especially if, as is intended, we try to understand it in a universal modal sense. No concrete physical "matter" exists other than as a typical-individual entity. The closest one can come to a modal characteristic of any particle is in its rest mass. We view a particle as a typical individualization of energy and with each typical particle structure there corresponds a typical rest mass.

There is at present no theory available which can account for the value of the rest masses of the known particles. Some features may rightly be called mysterious (for instance, the electron and the muon differ in nothing but their rest mass besides some properties such as their magnetic moment which depend on the different rest masses). However, there is no need to doubt that at some time in the future it will be possible to calculate the rest mass of any particle from typical fundamental interactions. A justification for this expectation can be found in quantum electrodynamics, in which the "renormalization procedure" can already partly account for the rest mass of the electron.

Particles are not merely characterized by their masses, but by sets of typical "quantum numbers", which allow us to classify particles in a few families. Although these properties are only partially understood, they have one feature in common, that allows us to distinguish particles from enkaptic wholes. In order to see this we must consider once again the concepts of superposition and superselection.

In quantum physics the state of a system is the superposition of the eigenvectors of the operators that describe the structure of the system. We may say that the system is "simultaneously" in all those eigenstates for which it has a non-zero probability, as found in a measuring procedure. Important properties which are subjected to this superposition rule are energy, position, linear momentum, and one component of angular momentum or spin.

As observed already in Sec. 9.7, there are physical properties which are not subjected to the superposition principle, although they too display a spectrum. These properties include the electric charge ($\pm e$), total spin (some integer or half-integer non-negative number), isospin (with different eigenvalues for protons and neutrons, e.g.), baryon number, and lepton number. They are subjected to super-selection rules. Thus there are positive and negative electrons, but there are no mixed states for electrons and positrons. There are no mixed states for any particles with integer and half-integer spin values.

This provides us with the sought for criterion for distinguishing particles as numerically founded structures from systems which are spatially founded (enkaptic wholes). Numerically founded structures are characterized by numerical properties subjected to superselection rules, whereas enkaptic wholes have internal degrees of freedom which are subjected to the superposition principle.

These considerations apply to the internal state only and not to external states. For electrons, for example, there are external parameters such as the kinetic energy, linear momentum, and one component of the spin, which are superposable. The spin component can have only two eigenvalues ($\pm \frac{1}{2}$) and therefore it has sometimes been assumed that the electron, too, has some kind of internal structure. Although this is very well possible, such a conclusion does not follow from the discrete eigenvalue spectrum for the spin, since the spin has to do with the external spatial orientation (with respect to a magnetic field), just as the kinetic energy and the linear momentum are related to the spatial and temporal homogeneity. Moreover, the two spin eigenstates can be superposed. The external superposable properties of electrons allow them to be enkaptically bound in atoms etc.

The typical properties which characterize the electron as a structure of individuality, viz. its charge, lepton number, and total spin value are not superposable with the values characterizing other particles of the same family. This means, for instance, that we cannot change an electron into a positron. Therefore, to identify the two particles merely as different states of the same system can only be done by analogical reasoning, although doing this has some obvious merits.

In contrast, enkaptic wholes such as molecules, atoms, and solids have internal properties (such as energy) which typically display a spectrum of superposable eigenvectors. The operators which determine their structure, among which the Hamiltonian is most important, all have this character. As a result, these structures can be described with the help of a Hilbert space, which is not the case with particles. Thus, on the subject side, the individuality of particles is mainly external, whereas, for enkaptic wholes, the individuality is also internal, which implies that we can speak of "the same system" whether it be in its ground state or in some excited state.

Thus we can speak of "particles" for two reasons. The first, from a retrocipatory viewpoint, is that particles can exist more or less freely and independently of other systems and are characterized by non-superposable properties. Secondly, from an anticipatory viewpoint, we can speak of particles if they retain these properties when enkaptically bound within spatially founded structures such as a nucleus, an atom, a molecule or a crystal. Quasi-particles only satisfy the first criterion. The other mentioned structures satisfy both criteria – for example, a nucleus is a particle within the structure of a molecule. Therefore we apply criteria of a typical kind and not of a modal kind. We thus reject the classical criterion of a particle as something localizable in space and time and endowed with mass.

10.6 *The opening up of the concept of a particle*
We have now completed our consideration of the concept of a particle in its retrocipatory thing-like structure. However, we have emphasized that a merely retrocipatory description is insufficient for a full understanding of any typical structure. A complete description must include the anticipatory direction, and such an opening up can be considered along several different lines.

The first of these is the purely modal one since any particle has energy, momentum and spin which can be displayed in several kinds of interaction. Next we could consider the particle's typical interaction with other particles, enkaptic wholes or quasi particles. We have already mentioned the collision cross section which clearly has an event-like, not a thing-like, structure.

In this same context we could also study the particle's stability. At high relative energies, no particle appears to be stable, but at relatively low energy levels, some particles can be distinguished from others by their stability. However, according to modern theories, even this stability must be considered the net result of a continuous creation and annihilation of virtually existing intermediary states. Thus a proton can be understood as a neutron with a cloud of mesons and vice versa. An electron is viewed as continually interacting with the electromagnetic field, during which virtual photons are emitted and absorbed, and electron-positron pairs are continually created and annihilated. Perhaps this language is merely metaphorical, since each of these processes (visualized in "Feynman diagrams") corresponds

210

to a term in a perturbation series, of which perhaps only the sum has any physical meaning.[14] Nevertheless, the view that an electron is a stable unchangeable thing is now rather suspect, especially if its event-structure is taken into account.

We shall consider yet another way in which the concept of a particle can be opened up, viz. in a typical way. We have stated that particles are numerically founded structures, implying that they can at least be taken together in a collection and counted. As a result, we must pay some attention to statistics. Because we are dealing with physically qualified structures, this "taking together" must be understood in a physical way. Either the collection of particles must be enclosed within more or less rigid walls (ultimately a potential well), or they must be enkaptically bound within a typical enkaptic whole. For our present purposes, these two cases are quite similar.

Since the spatial character of the enclosure is obvious and the possible motions of the particles become bounded (meaning that the possible kinetic energy and momentum states become superpositions of energy and momentum eigenstates respectively, with a discrete spectrum of values), both cases show the opening up of the thing-like character of a particle into an event-like character. This implies that we can count, not only the particles, but also the eigenstates, and we can study the distribution of the particles taken together over the possible eigenstates of the system.

Classical physics already utilized the application of statistics to study gases as conglomerates of large numbers of molecules. But the application of statistics in quantum physics is different in two respects. First, we can make use of the fact that the particles bound together are "like" or "similar". (Often the unfortunate term "identical" is used.) That is, all the particles are subjected to the same law, which means that, except for their external relations, they are the same.[15] Secondly, in quantum physics, statistics can already be used when only two like systems are involved, e.g., in the case of the helium atom with its two electrons.

We shall now discuss an approximate treatment of these problems. This approximation is so good that there is hardly ever any reason to correct for it. The assumption is that if we want to describe the state of a system consisting of two or more like particles, we can do so by constructing it as the product of the states of the separate particles. This means that the individual existence of the particles within the system as a whole is taken for granted: in statisti-

14. Bunge B 268, 269.
15. cf. Jammer D 338; Jauch 275; Klein A.

cal terms, we treat the particles as independent (cf. Sec. 8.3). If we number the particles as 1 and 2, the state can be abbreviated as:

$$f(1, 2) = f'(1)f''(2)$$

where the primes indicate that we are dealing with different states for the two particles.

We can now apply the results found in Chapter 9 regarding the symmetry of the system. If the two particles are like, i.e., subjected to the same typical law, having the same typical structure, then the state $f(1, 2)$ should undergo no basic change if we exchange the two particles. This means that under the exchange of the two particles, $f(1, 2)$ should be multiplied by a phase factor (Sec. 9.2), which in this case can only be ± 1. Thus $f(1, 2) = \pm f(2, 1)$. This result follows from the fact that if we change the order of the particles a second time, we must recover the original state $f(1, 2)$.[16] The general state can now be written as

$$f(1, 2) = f'(1)f''(2) \pm f''(1)f'(2)$$

In the first case ($+$ sign), we speak of a symmetrical state function and the particles satisfying this rule are called "bosons". In the second case ($-$ sign), we speak of an antisymmetrical state function, and the corresponding particles are called "fermions". Here we meet another superselection rule, since superposition is impossible between symmetric and antisymmetric states. Thus the property of being a fermion or a boson is a particle property. Without change, this theory can be extended to cases concerning more than two particles.

For composite particles such as nuclei and atoms, the fermion or boson character is determined by the number of composing elementary particles. For elementary particles, it is found that particles having half integer spin (electrons, protons, neutrons, neutrinos) are fermions, whereas those having integer spin values (light quanta, sound quanta, mesons) are bosons. In fact, elementary bosons are not particles, but quasi-particles, i.e., individualizations of currents.

This state of affairs is particularly important for fermions, since, as is readily seen, it implies that no two particles of the same kind can be in the same state within the same system.[17] If the states distinguished by primes would be the same, $f(1, 2)$ would become zero. This statement (Pauli's Exclusion Principle) is meaningful only if the

16. Expressed in group-theoretical terms, there are two "permutation groups", the symmetrical and the anti-symmetrical group.
17. See, e.g., Margenau A, Ch. 20; Weisskopf 200: "In many ways, the Pauli principle replaces the classical concept of impenetrability or hardness. Two identical particles, obeying the principle, can never be brought to the same place. It is therefore reasonable to reserve the term *particle* for the entities that obey the Pauli principle".

two states distinguished by primes are distinguishable by quantum numbers. Thus it is not necessary that the system containing the particles form an enkaptic whole for even in a box with impenetrable walls or in a potential well the eigenstates of the enclosed particles have a discrete spectrum.

In this case we can speak of an anticipation of spatially founded structures, because in this statistical analysis it is assumed that the particles are simultaneously present. It is also assumed that they retain their own typical properties and their own individual existence, at least if they are fermions or composite bosons. Contrasted with this behaviour is that of quasi-particles, which lose their individual existence after they are absorbed into a spatially founded typical whole. This does not mean that fermions retain their individual existence forever. The collision of two electrons, for instance, must be described in such a way that both lose their identity.[18] Thus four different electrons are involved in such a process, two before and two after the collision. This statement has relevant meaning which can be checked experimentally. It can be compared with the collision of two droplets of water. If, afterwards, two new droplets are formed, we are dealing with a similar "four-particle interaction".

The commonly held view that the theories discussed above lead to the rejection of the particle's individual existence rests upon the confusion of "individuality" with "identifiability" in a modal sense. Classical physics assumed that, in principle, any particle could be identified at any time by its momentum and position. This notion, for example, is the starting point of the Maxwell-Boltzmann distribution, which differs sharply from the quantum mechanical Bose-Einstein and Fermi-Dirac distributions. Quantum physics has shown that such identifiability is impossible. Especially, if two or more like particles interact within some system, it is, in general, impossible to identify them. Only if, in some energy region, the number of particles is far less than the number of available states (if the occupation number is much less than unity), do the Bose-Einstein and Fermi-Dirac statistics approximate the classical Maxwell-Boltzmann distribution.

This state of affairs has often led to the denial of the individuality of those particles. Others, however, have correctly countered this argument by pointing to the fact that these particles can still be counted,[19] still have relative positions and motions and still interact with each other. In order to account for these subject-subject re-

18. Jammer D 344.
19. Harris A 135; Jammer D 345; Margenau A 441.

lations, the separate individual existence of the particles must be assumed. In classical physics, on the other hand, the identifiability of particles rests on the presumed determinate character of their motions and interactions, a supposition which is inconsistent with the individuality of the particles and which is untenable in quantum physics.[20]

In our view, individuality means that, in the law-subject relation, the individual behaviour of particles is not determined, but is at most delimited by their typical structure and the structure of the systems with which they interact. In this way, we can still speak of the "likeness" of particles (subjected to the same typical law), *and* retain their individual differences.

10.7 *Spatially founded structures of individuality*

If two or more physically qualified systems (particles or otherwise) combine to form a new system, several different cases can be distinguished. In the first case, one or more of the combining systems may completely lose its individual existence. This process is called "absorption" and is exemplified by light quanta being absorbed by an atom or an electron being absorbed by a nucleus. In a second case, the combining systems do not form a new structural whole, but form a mixture or an aggregate, often because of external boundaries. Gases in a container, unordered alloys of metals in a solid, liquids, etc., must be considered mixtures in this sense. In this instance, the combining systems completely retain their individual existence.

Dooyeweerd[21] has introduced the term "enkapsis" or "interlacement" to describe a third type of combination, more or less intermediate between the first two cases. We speak of enkapsis if two or more systems, each with its own typical structure, combine to form a new system, characterized by its own distinct typical structure, but in which the composing systems retain their own structure and individual existence.[22] Thus, protons and neutrons combine to form nuclei, a nucleus together with one or more electrons form an atom, two or more atoms may combine to form a molecule, and a large number of atoms or molecules combine to form solid crystals. In each of these cases, the protons, neutrons, electrons, atoms and molecules are enkaptically bound within a new structural whole, but they retain their own individual existence and typical structure.

20. Jammer D 344; Jauch 276.
21. Dooyeweerd D 694-713 discusses the enkaptic relation between an atom and its nucleus and electrons.
22. See also Harris A 136, who speaks of ". . . structured totalities, which are neither simple unities nor dissectable aggregates, but are diversified wholes of distinguishable though inseparable constituents".

214

This state of affairs does not mean that the structures of enkaptically bound systems remain completely unchanged. For example, the structure of a free atom is not the same as the structure of that atom when it is bound within a molecule, because the structure of any system is dependent, to some extent, on its environment. The symmetry of a free atom is different from that of the same atom placed in a magnetic field, and if this magnetic field is provided by the structural whole (the molecule) in which the atom becomes enkaptically bound, its structure will be changed to some extent. Similarly, in a nucleus, protons and neutrons acquire an interdependent existence, and by exchanging pions, protons become neutrons and vice versa.

However, such changes are merely slight adaptations of what is already implicit in the structure. If, for example, we find that in the presence of an external field a single energy level is split into two energy levels, this implies that the original energy level was already degenerate. Thus even in the absence of the external field, the energy level is already a double level, although the two levels have the same energy eigenvalue. It is generally true that a system undergoes a change in symmetry when it is enkaptically bound within a new structure.

Those systems which combine in the process of absorption are related to kinematically founded systems, since the absorption concerns individualized currents. Those in which combination results in aggregates or mixtures, are related to numerically founded structures, since this combination concerns the mere addition of particles. Those cases which involve enkapsis are always related to spatially founded structures, since in the new structure, the structures of the composing systems always exist simultaneously. The recognition of these three basic types of combination is crucial for a good insight into some of the fundamental problems of modern physics.

The typical law for an enkaptic structural whole is objectively described by a Hamiltonian operator in which the kinetic energy of the composing particles, the potential energy (which determines the typicality of the structure), and the so-called "exchange energy" are represented. The electrons in an atom, for example, tend to avoid each other ("avoid" in a broad, nor merely spatial, sense) because of their fermion character, and thereby create a kind of "repelling force". Bosons, on the other hand, tend to attract each other.[23] At relatively low temperatures this "exchange interaction" is displayed

23. Margenau A 434.

in widely differing phenomena[24] like the superfluidity of liquid helium, the superconductivity of many metals, the temperature dependent specific heat of solids, ferromagnetism, etc. The chemical properties of atoms and molecules are also highly influenced by the fermion character of the electrons bound in these systems. In fact, the periodic system of the elements must be explained with the help of Pauli's exclusion principle. The interaction between nucleons in a nucleus too is determined mostly by exchange interaction.

It is unusual to treat "exchange interaction" on the same footing as electromagnetic interaction or one of the other fundamental interactions. In the latter case the interaction arises from a typical physical property of the interacting particles, such that the interaction itself and its retrocipatory analogies are typical. In exchange interaction the similarity of the subjects directly affects their mathematical modal subject-subject relations functioning in opened-up form – i.e., exchange anticipates physical interaction. For instance, in a helium atom, consisting of a nucleus and two electrons, the electromagnetic interaction depends on the mutual distance of the electrons. The exchange relation produces "exchange energy" by influencing this mutual distance, and thus indirectly affecting the interactions of the electrons and the nucleus.

Not every physically qualified ensemble of particles is a typical whole. A gas, e.g., is a mixture, an aggregate of freely moving, though interacting (colliding) molecules. We can also attribute some persistence, coherence, and internal equilibrium to a mixture. But we shall show presently that these properties are of an external origin and have to be distinguished from the *typical* permanence, coherence, and stability of individual wholes. Generally speaking, they depend on the boundary conditions for a mixture, whereas they are largely independent of external influences for a typical whole.

We have seen that the persistence of a physical subject is related to its total internal energy or rest mass which is conserved as long as the subject does not interact physically with other subjects. We may observe that the mass of a typical whole (or, in the case of a crystal, its mass density at constant temperature and pressure) has a typical numerical value. All hydrogen atoms, e.g., have the same mass. But the energy of a mixture does not have a typical value. In particular the internal energy of a gas, being dependent on the number of molecules, the temperature and volume of the container, may have any value within a wide range.

24. Jammer D 145, 342, 343.

Furthermore, the notion of a mixture implies a certain non-typical coherence, usually determined by external spatial boundaries – e.g., the walls of a vessel. Both the subjective duration and extension of a mixture are determined by these external boundaries. But the extension of a typical whole is determined by its internal structure. For example, the interatomic distance is characteristic of the structure of a crystal.

Mixtures lack the internal symmetry which is characteristic of a typical coherent whole. A solid solution of a metal in another metal is usually an aggregate. Since these alloys, with only few exceptions, have lower melting points than the pure metal, they are less stable. Only if two metals form an ordered structure (an intermetallic compound) with a fixed, usually simple ratio between the numbers of atoms of the composing metals may this alloy have a melting point higher than that of the pure metals, or at least higher than alloys with slightly different compositions.

We also have to distinguish the stability of a typical physically qualified whole from an equilibrium state of a mixture. The latter state is a non-typical relation between parts of the system and is determined by the uniformity of temperature and other intensive parameters (cf. Chapter 5). Nothing of this kind applies to a typical whole for which any notion of uniformity is alien. The internal stability of a typical whole like a crystal is determined by a typical balance (determined by a typical law) of kinetic energy, potential energy, and exchange energy. On the other hand, the equilibrium of a mixture like a gas comes about by a random distribution of energy over the composing particles. Furthermore, the motions of the particles are random, whereas the motions of enkaptically bound particles in a physically qualified enkaptic whole are restricted to typical periodic motions.

This does not mean that there is no modal criterion for the stability of both mixtures and typical wholes. Utilizing the First and Second Laws of thermodynamics one can prove that the most stable state for an isolated system is the one of lowest energy, provided there are two possible states (according to structural laws) with different energies. In thermal physics one is often more interested in systems which are in thermal equilibrium with a thermostat or in systems which are kept at constant pressure than in isolated systems. In the first case, the most stable state is the one with the lowest value for the "free energy". In the constant pressure case, the state of lowest "free enthalpy" is the most stable. In this manner one can give modal, thermodynamic criteria for phase transitions like that between a gas and a liquid, or that between the superconducting and the normal state of a metal. Although one still needs a structural

217

analysis of the phenomena in all these cases, the phenomena are subjected to modal laws of stability.

10.8 *Quasi-particles*

Particles can interact with each other by exerting forces. If the potential energy has a minimum with respect to the relative positions of the particles they can form an enkaptic whole provided their relative kinetic energy is not much larger than the depth of the potential minimum. Enkaptic wholes can also interact with each other in this way, although the forces are often more complicated in this case. The Van der Waals interaction and the ionic, covalent, or metallic binding in solids are all supposed to be reducible to electromagnetic interactions, but are nevertheless still not completely under control.

Particles and enkaptic wholes are not restricted to static interactions via forces, but can also interact dynamically via currents. In order to account for currents the retrociprocatory spatial analogy of interaction (force) must first be opened up into the concept of a field, and, as we saw in Chapter 7, opened-up motion in a field has the character of wave motion. Hence the wave packets representing currents in a field (or, eventually, in a solid) are kinematically founded individualizations of these currents. They are generally called quasi-particles.

In solid state physics, the typical character, motion, emission and absorption of phonons, magnons, polarons, etc., are fairly well understood. To say that a crystal contains quasi-particles is equivalent to saying that the system is in an excited state. Therefore, at times, one calls them "excitons" and the number of excitons in a certain state is called the latter's occupation number. In electromagnetic fields the interaction of photons with atoms or electrons is studied in quantum electrodynamics. We shall only make a few remarks about quasi-particles.

First we observe that quasi-particles cannot be considered as "building blocks" of enkaptic wholes, although they may constitute the indispensable "glue". If an electron is enkaptically bound together with a proton to form an hydrogen atom, its binding energy is 13.6 eV, which is far less than its total energy of 0.51 MeV (equivalent to its rest mass). On the other hand, the binding energy of a pion in a nucleon is its full 140 MeV rest energy. If an electron is removed from an hydrogen atom, we are left with a proton. If a pion is emitted by a nucleon, we still have a nucleon. Similarly, atoms can absorb or emit photons.

A photon has no individual existence within the atomic structure, but only when moving between its emission in one atom and its absorption in a second atom. Since these "things" cannot be foundational to enkaptic wholes, we, therefore, speak of "quasi-particles". Because their foundational modal aspect is the kinematic one, which immediately precedes the physical, the retrocipatory, thing-like character of quasi-particles is less pronounced than their anticipatory, event-like structure. Emission and absorption is more relevant for quasi-particles than for particles. Particles, however, may also undergo total absorption or emission. An electron, for example, cannot exist independently in a nucleus.

Secondly, force, as a retrocipatory spatial analogy of interaction and having static character, has no direct application to quasi-particles. However, its opened-up counterpart, a field, is relevant. Also, momentum can be attributed to quasi-particles. This implies that Newton's third law (action = − reaction) is not applicable to interactions in which individualized currents are involved, whereas the conservation law of linear momentum can be applied to any isolated system, whether or not it encloses individualized currents. Quasi-particles are currents in a real field of interaction (electromagnetic or otherwise). Particles also show wave motion, but now in an imaginary field (with complex coefficients, cp. Sec. 7.4). The latter field points to a spatial anticipation on the physical modal aspect, whereas a real field shows the opening up of a spatial retrocipation of physical interaction, i.e., force. The two cases are very similar, but can be distinguished.

Finally, we find that quasi-particles are invariably bosons, and have integer spin values, whereas elementary particles are fermions with half integer spin.[25] This means that quasi-particles are not subjected to Pauli's exclusion principle. Thus, if they are enclosed in some spatial region (a vessel or a solid crystal), there can be an unlimited number of quasi-particles in any state of that system, and this number is adjusted so as to satisfy the equilibrium conditions dictated by the environment. E.g., in the radiation enclosed in a black box the number of photons and their distribution over all allowed frequency states are completely determined by the volume of the box and the temperature of its walls. If one of these parameters changes, the number of photons in each state is "automatically" adjusted.

In contrast, the number of electrons in a solid crystal is independent of temperature and volume (if the number of atoms is kept constant). Hence, whereas enkaptic wholes are based on particles, quasi-particles exist on the foundation of enkaptic wholes, and can be understood on the basis of the opening up of the event-structure of the latter. It should be observed that the number of electrons is not constant in all

25. Akhieser, Berestetsky 524-527.

circumstances. A high energy photon may give rise to the spontaneous creation of an electron-positron pair, and in solid state physics, a phonon or a photon may create an electron-hole pair (in several respects such a pair can be considered an exciton). Therefore we have to relax the requirement of constancy in the number of electrons and find the conservation of "lepton number". In any known physical process the number of leptons (electrons, muons, neutrinos) is constant, only if one counts the "antiparticles" (positrons, antimuons, antineutrinos) negatively. Similarly, there is a conservation law for baryons (protons, neutrons, hyperons, and their antiparticles). But no such law exists for any kind of bosons.

10.9 *Classical and modern physics*

Time and again three epochs of physical research are distinguished: *antique physics*, started by the Ionian philosophers of nature, but dominated by Aristotle until the end of the Middle Ages; *classical physics*, initiated by Copernicus and Galileo, dominated by Newton's mechanics, but also including classical chemistry, thermodynamics, optics, classical electromagnetism, and statistical physics; and *modern physics*, ushered in by Planck, dominated by Einstein and Bohr, and comprised of relativity theory and quantum physics.[26]

Classical physics may be distinguished from antique physics in various ways, for instance: the introduction of experiments as means to verify theories; mathematical objectification; abstraction – the modal (especially spatial and kinematic) description of physical states of affairs; the search for universality (heavenly and earthly motions considered to be subjected to the same laws). In Aristotelian physics the typicality of everything was most important, whereas in classical physics the kinematic modal aspect as universal mode of explanation became the leading motive.

The distinction between classical and modern physics is also manifold. In particular quantum physics has changed the face of physics in five different, though interdependent ways.

(*a*) *The qualifying modal aspect* – In classical physics it was assumed that ". . . all change and especially all qualitative change, has to be explained by the spatial movement of unchanging bits of matter – by atoms moving in the void".[27]

26. Jaki, Ch. 1-3. This is a rather crude distinction. For example, many modern insights of Galileo's can be traced back to medieval kinematics, and even after Galileo there was still much Aristotelian thinking; cf. Truesdell 30-32, 89; Butterfield, Ch. 1; Basalla; Finocchiaro.

27. Popper B 146. See also Capek. Ch. 5, 6; Dijksterhuis 456, 474ff; Einstein B 18ff; Heisenberg C 52; Jaki, Ch. 2; Jammer B 103ff, 221ff, C 90; Jauch 68ff; Whitrow 3.

In modern physics the qualifying aspect is no longer dynamic motion, but the irreducible modal aspect of physical action.[28] In the 19th century, the mechanicist view was already challenged because of some inconsistencies with respect to thermal physics and electromagnetism.[29] The "energeticists" (Ostwald, Mach, Duhem) recognized the necessity of breaking with this view, but were overruled, perhaps mainly because they wanted to replace one absolute (mechanicism) by another one (energeticism). One of their tenets was that atoms and molecules were merely theoretical constructs which should be banished from the empirical science of physics. Hence energeticism lost much of its appeal in the early years of the 20th century when the existence of atoms and molecules became established beyond any reasonable doubt.

However, the mechanicist atomic theory also received its death blow at nearly the same time – not because atoms turned out to be composite systems, but because motion could no longer be considered as the main principle of explanation, i.e., the leading, qualifying modal aspect. Time and again, one has tried to save the old theory by introducing new, more "elementary" particles on the one hand, and a "unified field theory" on the other hand, so that the old view of unchangeable particles moving in a field could stay in the all-embracing model. These attempts have failed on both sides. Hardly anyone today believes that there are really elementary building blocks of matter, and that all kinds of interaction can be reduced to a single one. The dualism of matter and force seems to be history by now.

(b) The opening up of the pre-physical modal aspects – The absolutistic mechanicist view that motion is the qualifying modal aspect for physical states of affairs precluded the opening up of the kinematic modal aspect, but also of the numerical and spatial modal aspects. One of the most striking features of modern physics is the development of these anticipations: the kinematic analogies in the numerical and spatial modal aspects by relativity theory (see Chapter 4), and the physical analogies in the pre-physical modal aspects by quantum theory. The latter became possible only when the physical modal aspect was implicity accepted as irreducible to the kinematical one. This development is most pronounced in the theory of wave motion (see Chapter 7), whose foundations were laid down by Huygens, Fresnel, Maxwell, and others. This theory did not receive a universal (modal) status until De Broglie's hypothesis of electron waves.

28. Cantore 270f. Not every physicist has broken with the ideal of describing physical states of affairs in kinematic terms: cf. Heisenberg C 43; Berestetsky, in: Kuznetsov 26; Omelyanowski, in: Kuznetsov 237; Margenau A 429.
29. Čapek, Ch. 7.

Time and again we have stressed the relational character of the modal aspects, which are properly called aspects of time. In relativity theory, absolute concepts like simultaneity and mass turned out to be frame dependent. The conservation laws were proved to be a most important consequence of the relational character of mutually irreducable modal aspects.

(c) *The recognition of individuality in physics, and the acceptance of statistical methods* – We have seen (Chapter 8) that individuality and probability, to a very large extent, denote the subject and law side, respectively, on the typical side of temporal reality. Statistical methods were introduced in the 19th century by Maxwell and Boltzmann for want of anything better, since modal kinematic determinism was still accepted as an article of faith. The discovery of radioactivity, the explanation of black-body radiation, Brownian motion, and photoelectricity, and the rise of quantum physics, mark the entrance of individuality in physical thought. We showed that probability is related to the anticipatory direction on the typical side of reality. Its applicability in modern physics depends, therefore, on the development described under (b).

(d) *The concept of a state* – As a spatial objective representation of a physical system the concept of a state was already applied in thermodynamics, but not in all its consequences. In particular the spatial superposition principle had no meaning in thermal physics. After the introduction of the main principles of quantum physics, physicists realized that the state of a system can be represented by a vector in a complex Hilbert space so that the formalism is able to account for the points mentioned above (see Chapter 9).

(e) *The distinction of modality and typicality* (both on the subject side and the law side).[30] – We already mentioned the implication of this distinction for individuality and probability. In addition it was recognized that the typical interactions (like the electromagnetic one) cannot be reduced to modal relations, and, therefore, that the typical structures of individuality (like that of an atom) cannot be fully analyzed in modal concepts, let alone in dynamic ones. This development again depends on those mentioned above. In fact, this point marks the most significant breakthrough in the physics of our century, because it enabled the study of atomic, molecular, crystalline and nuclear structures, and, more recently, also of subnuclear and stellar structures. At first, quantum theory was mainly concerned with statistical problems (Planck, Einstein), but the theory received its major impetus from Bohr's central problem: Why is an atom stable? We discussed this and related problems in the present chapter.

30. Cantore 266-280.

Time and again other differences between classical and modern physics are mentioned, which must be considered less relevant, obscurely stated, or even false. We shall recall three of them.

First, one often finds the view that quantum physics is only relevant on a microscopic scale. Classical physics is assumed to still be valid on a macroscopic scale. It may be true that we can still use classical physics to describe the *motion* of macroscopic bodies, but the distinction of classical and modern physics is far more involved. In particular solid state physics, plasma physics and astrophysics are completely beyond the reach of classical physics, and certainly do not remain on a microscopic level. This view often comes up in discussions of measurement theory, but it is overlooked that there are many observable macroscopic effects which cannot be accounted for by any classical theory.

Secondly, it is customary to consider classical physics as a limiting case of quantum physics. Again, this is right with respect to the kinematics of physical systems, but it does not do full justice to classical physics. In our view, the latter was wrong in its attempt to explain everything in modal terms, especially in modal kinematic terms. But the modal analysis performed in classical physics keeps its irreplaceable value: it is simply not sufficient. Therefore, in our systematic analysis classical physics is rated higher than in the usual view which considers it as a boundary case. This applies to classical mechanics and thermal physics, as well as to classical electromagnetism.

Thirdly, we consider it a fallacy to state that quantum physics is less objective than classical physics. This view is inspired by the fact that observation of a physical system in modern physics involves some interaction between the observer and the observed object. In classical physics scientists assumed this interaction could be made arbitrarily small. We have seen (Sec. 1.6) that the customary distinction of subject and object is highly influenced by a philosophy which cannot be ours. Observation is strictly speaking not a physical act, but a psychical one which has a physical aspect. In physics we ought to be interested only in the latter, and we can describe all physical states of affairs as objectively as we wish – provided that the meaning of the term "objective" has lost its traditional obscurities. Therefore, we do not deal with observers, but instead discuss the relevance of interaction for the concept of a state as an objective representation of a physical system.

By and large, we can say that the most striking aspect of 20th-century physics is the emphasis on the analysis of the structure of matter. As observed, classical physics turned its back to this analysis, in favour of a purely modal explanation of natural phenomena. Of course, this does not mean that modern physics implies a return to Aristotelian physics. In the present chapter we have tried to show that the modal analysis, mostly due to classical physics, is a necessary prerequisite for a fruitful study of typical structures.

Apart from the Philosophy of the Cosmonomic Idea, no contemporary philosophy is able to give a systematic analysis of physically qualified structures of individuality as thoroughly as was given in the present chapter. The exploration of the mutually irreducible modal aspects as principles of explanation and of temporal order was undertaken without making use of any of the results of the study of the typical structures, and it could not be expected to have this result.

Time and again we stressed the complementary character of our systematics. It is not conducted by a single principle of explanation, but by the mutually orthogonal, but complementary ideas of law and subject, modality and typicality, and of the various modal aspects. With their help we have tried to lay bare the philosophical foundations of the physical sciences.

Name Index and Bibliography

Agassi, J.
- (A) *Faraday as a Natural Philosopher*, Chicago, 1971 – 110, 111
- (B) Towards an Historiography of Science, in: *History and Theory*, Beiheft 2, 1963 – 23

Akhieser A. I., Berestetsky V. B.
- *Quantum Electrodynamics*, Oak Ridge, Tenn., 1953 – 95, 149, 219

Alembert, J. B. d' – 111

Aristotle – 3, 20, 33, 34, 80, 82, 113, 220, 224

Avogadro, A. – 197

Ballentine, L. E. – 184, 185
- The Statistical Interpretation of Quantum Mechanics, in: *Reviews of Modern Physics* 42, 1970, 358-381 – 184

Bar-Hillel, Y. – see Fraenkel

Basalla, G.
- (ed.), *The Rise of Modern Science*, Lexington, Mass., 1968 – 220

Bastin, T.
- (ed.), *Quantum Theory and Beyond*, Cambridge, 1971 – 151, 184, 193

Berestetsky, V. B. – 221 – see also Akhieser

Bergmann, P. G.
- The Special Theory of Relativity; The General Theory of Relativity, in: *Handbuch der Physik*, vol. 4, Berlin, 1962, 109-202, 203-272 – 95

Berkeley, G. – 54

Berkson, W.
- *Fields of Force*, London, 1974 – 110, 111

Bernard, C. – 30, 153

Bernoulli, J. – 159

Berzelius. J. J. – 149

Beth, E. W. – 50
- (A) *De Wijsbegeerte der Wiskunde van Parmenides tot Bolzano*, Antwerpen, 1944 – 33, 34, 41, 42, 136
- (B) *Wijsbegeerte der Wiskunde*, Antwerpen, 1948 – 40-42, 44, 45
- (C) *Natuurphilosophie*, Gorinchem, 1948 – 4, 88
- (D) *Wijsgerige Ruimteleer*, Antwerpen, 1950 – 42, 47, 55, 75

Bohm. D. – 150, 193
- *Causality and Chance in Modern Physics* (1957), London, 1967 – 150

Bohr. N. – 24, 76, 151, 177, 193, 195, 196, 220, 222
- (A) Discussion with Einstein on Epistemological Problems in Atomic Physics in: Schilpp A 199-241 (also in Bohr C) – 25, 145, 146, 151, 177, 185
- (B) *Atomic Physics and the Description of Nature* (1934), Cambridge, 1961 – 150, 177
- (C) *Atomic Physics and Human Knowledge*, New York, 1958 – 177
- (D) *Essays 1958-1962 on Atomic Physics and Human Knowledge*, New York, 1963 – 177

Boltzmann, L. E. – 115, 118-120, 129, 160-163, 165-167, 192, 207, 213, 222
Bondi, H. – 129
Boole, G. – 156, 174
 – *An Investigation of the Laws of Thought* (1854), New York, 1958 – 156
Borel, E. – 156
Born, M. – 164, 169
 – *Natural Philosophy of Cause and Chance* (1949), New York, 1964 – 71, 160, 161
Boscovich, R. G. – 110
Bose, S., 163, 213
Boyer, C. B.
 – *The History of the Calculus and its Conceptual Development* (1939), New York, 1959 – 40
Boyle, R. – 161
Brahe, T. – 81, 82
Braithwaite, R. B.
 – *Scientific Explanation* (1953), Cambridge, 1968 – 9, 130, 158
Bridgman, P. W. – 87, 201
 – *The Logic of Modern Physics* (1927), New York, 1954 – 66, 196
Brillouin, L.
 – *Scientific Uncertainty and Information*, New York, 1964 – 127, 129
Brody, B. A.
 – *Readings in the Philosophy of Science*, Englewood Cliffs, N. J., 1970
Broglie, L. V. de – 135, 141, 142, 146, 150, 193, 221
Brown, R. – 119, 120, 122, 149, 154, 197, 222
Büchel, W.
 – Zur Begründung und Deutung der Relativitätstheorie, in: *Philosophia Naturalis* 10, 1968, 211-236 – 97
Bunge, M. – 2-4, 13
 – (A) *Causality* (1959), Cleveland, 1963 – 8, 90, 101, 130, 131
 – (B) *Foundations of Physics*, Berlin, 1967 – 2-4, 8, 10, 13, 60, 70, 88-90, 92, 95-98, 108, 142, 149, 155, 177, 211
 – (C) *Scientific Research*, vol. 1, Berlin 1967 – 66, 96
 – (D) *ibid.*, vol. 2 – 60, 62, 65, 66, 68-70
 – (E) Physical Axiomatics, in: *Rev. Mod. Phys.* 39, 1967, 463-474 – 2, 4, 102
 – (F) (ed.), *Quantum Theory and Reality*, Berlin, 1967
 – (G) (ed.), *The Critical Approach to Science and Philosophy*, London, 1964
 – (H) (ed.), *Delaware Seminar in the Foundations of Physics*, Berlin, 1967 – 2
 – (I) *Metascientific Queries*, Springfield, III., 1959 – 177
Butterfield, H.
 – *The Origins of Modern Science* (1949), London, 1968 – 80, 120
Byerly, H. C., Lazapa, V. A.
 – Realist Foundations of Measurement, in: *Philosophy of Science*, 40, 1973, 10-28 – 66
Callen, H. B.
 – *Thermodynamics*, New York, 1960 – 104
Campbell, N. – 17
 – (A) *What is Science?* (1921), New York, 1952 – 13, 60, 61, 69, 79, 130, 131
 – (B) *An Account of the Principles of Measurement and Calculation*, London, 1928 – 60, 61, 65, 66, 69
Cantor, G. F. L. P. – 40
Cantore, E. – 3
 – *Atomic Order*, Cambridge, Mass., 1969 – 5, 196, 198, 207, 221, 222

Čapek, M.
- *The Philosophical Impact of Contemporary Physics*, Princeton, N. J., 1961 –
50, 53, 80, 149, 151, 153, 220, 221
Carathéodory, C. – 71, 102
Carnap, R. – 59, 158
- (A) *Foundations of Logic and Mathematics* (1939), Chicago, 1970 – 35
- (B) *Logical Foundations of Probability*, Chicago, 1950 – 59, 158
- (C) *Philosophical Foundations of Physics*, New York, 1966 – 66
Carnot, N. L. S. – 72, 102
Cassirer, E.
- *Substance and Function* (1910); *Einstein's Theory of Relativity* (1921), New
York, 1953 – 33, 37, 66
Cauchy, A. L. – 40
Celsius, A. – 64
Čerenkov, P. A. – 92
Chissick, S. S. – see Price
Churchman, C. W., Ratoosh, P.
- (eds.), *Measurement, Definitions and Theories*, New York, 1959
Clausius, R. L. E. – 71, 72, 102
Cohen, L. – see Margenau
Colodny, R. G. – see Feyerabend C. D.
- (ed.), *Paradigms and Paradoxes*, Pittsburgh, 1972
Compton, A. H. – 185
Copernicus, N. – 81, 82, 220
- *On the Revolutions of the Heavenly Spheres* (1543), Newton Abbot, 1976 –
81, 82
Coriolis, C. G. de – 95, 96
Coulomb, C. A. de – 11, 112, 135
Courant, R.
- *Differential and Integral Calculus*, vol. 1 (1934), London, 1955 – 38, 40
Curie, P. – 98, 106
Dalton, J. – 149
Dam, H. van – see Houtappel
Danto, A., Morgenbesser, S.
- (eds.), *Philosophy of Science* (1960), Cleveland, Ohio, 1969
Davisson, C. J. – 146
Dedekind, J. W. R. – 40
Democritus – 206
Descartes, R. – 6, 19, 48, 51, 52, 54, 55, 107, 207
DeWitt, B. S., Graham, R. N.
- Resource Letter IQM-1 on the Interpretation of Quantum Mechanics, in:
American Journal of Physics 69, 1971, 724-738 – 150
Dirac, P. A. M. – 137, 163, 169, 181, 213
Dooyeweerd, H. – 4, 6, 14, 15, 25, 40, 214
- (A) Het Tijdsprobleem in de Wijsbegeerte der Wetsidee, in: *Philosophia
Reformata* 5, 1940, 160-182, 193-234 – 15, 35, 50
- (B) *A New Critique of Theoretical Thought* (1935 – 36), vol. 1, Amsterdam,
1953 – 4, 6, 9, 14, 15, 19, 20, 22, 25, 30, 130
- (C) *ibid.*, vol. 2, 1955 – 3, 4, 14, 15, 17-20, 22, 33, 35, 40, 46, 48, 50, 85, 130
- (D) *ibid.*, vol. 3, 1957 – 4, 18, 124, 214
- (E) *In the Twilight of Western Thought*, Nutley, N. J., 1965 – 4, 14, 19

Doran, B. G.
 – Origins and Consolidation of Field Theory in 19th Century Britain, in: *Hist. Stud. Phys. Sci.* 6, 1975, 133-260 – 88
Drake, S.
 – (ed.), *Discoveries and Opinions of Galileo*, Garden City, N.Y., 1957 – 21
Duhem, P. M. M. – 221
Dijksterhuis, E. J.
 – *De Mechanisering van het Wereldbeeld*, Amsterdam, 1950 – 80-83, 111, 206, 220
Eddington, A.
 – *Space, Time, and Gravitation* (1920), Cambridge, 1958 – 97
Einstein, A. – 23, 26, 27, 86-89, 93-95, 97, 98, 107, 121, 123, 150, 163, 178, 185, 186, 195-197, 207, 213, 220, 222
 – (A) On the Electrodynamics of Moving Bodies (1905); The Foundations of the General Theory of Relativity (1916), in: Lorentz et al., 35-65, 109-164 – 87, 93
 – (B) Autobiographical Notes; Reply to Criticisms, in: Schilpp A 1-95, 663-688 – 87, 148, 150, 220
 – Podolsky, B., Rosen N.
 Can Quantum-Mechanical Description of Physical Reality be Considered Complete? in: *Physical Review* 47, 1935, 777-780 – 185
Elkana, Y.
 – (A) Helmholtz' "Kraft", in: *Hist. St. Phys. Sci.* 2, 1970, 263-298 – 107
 (B) *The Discovery of the Conservation of Energy*, London, 1974 – 107
Ellis, B.
 – *Basic Concepts of Measurement*, Cambridge, 1966 – 60
Euclid – 2, 11, 26, 48, 50-53, 55, 56, 75, 89, 92, 94, 97, 98, 108, 170
Euler, L. – 107
Feigl, H. – 76
Feinberg, G.
 – Possibility of Faster-Than-Light Particles, in: *Phys. Rev.* 159, 1967, 1089-1105 – 91
Fermi, E. – 163, 213
Feyerabend, P. K. – 24, 25, 27, 29
 – (A) How to be a Good Empiricist (1963), in: Brody, 319-342 – 25, 27
 – (B) Realism and Instrumentalism, in: Bunge G 280-308 – 81
 – (C) Problems of Microphysics, in: R. G. Colodny (ed.), *Frontiers of Science and Philosophy* (1962), London, 1964, 189-283 – 81, 146, 177, 185, 191
 – (D) Problems of Empiricism, in: R. G. Colodny (ed.), *Beyond the Edge of Certainty*, Englewood Cliffs, N, J., 1965, 145-260 – 25, 27, 31
 – (E) *Against Method*, London, 1975 – 10, 25, 82
 – (F) *Science in a Free Society*, London, 1978 – 81
Feynman, R. Q. – 210
Fine, A.
 – Some Conceptual Problems of Quantum Theory, in: Colodny 3-31 – 146
Finocchiaro, M. A.
 – *History of Science as Explanation*, Detroit, 1973 – 10, 23, 220
Fizeau, A. H. L. – 144
Fourier, J. – 139
Fraenkel, A. A., Bar-Hillel, Y.
 – *Foundations of Set Theory*, Amsterdam, 1958 – 4
Frank, P.
 – *Modern Science and its Philosophy* (1941), New York, 1961 – 1, 130

Frege, G. – 33, 34
Fresnel, A. J. – 144, 221
Friedman, M.
– Grünbaum on the Conventionality of Geometry, in: *Synthese*, 24, 1972, 219-235 – 54
Frisch, O. R. – 184
Gale, M.
– (ed.), *The Philosophy of Time*, Garden City, N.Y., 1967 – 16, 84
Galileo Galilei – 11, 21, 26, 27, 80-82, 93, 108, 113, 140, 179, 181, 186, 220
– (A) *Dialogue Concerning the Two Chief World Systems* (1632), Berkeley, 1974 – 82
– (B) *Dialogues Concerning Two New Sciences* (1638), New York, 1954 – 80
Gauss, C. F. – 55, 56, 65
Germer, L. H. – 146
Gibbs, J. W. – 160, 162, 165, 167
Giles, R.
– *Mathematical Foundations of Thermodynamics*, Oxford, 1964 – 103
Gillispie, C. C.
– *The Edge of Objectivity* (1960), Princeton, N. J., 1973 – 80, 81
Gödel, K.
– *On Formally Undecidable Propositions of Principia Mathematica and Related Systems* (1931), Edinburgh, 1962 – 3
Gold, T.
– (ed.), *The Nature of Time*, Ithaca, N.Y., 1967 – 119, 129
Goldberg, S.
– In Defence of Ether, in: *Hist. Stud. Phys. Sci.* 2, 1970, 89-125 – 88
Goldstein, H.
– *Classical Mechanics* (1950), Reading, Mass., 1959 – 92, 105
Goodfield, J. – see Toulmin
Gordon, W. – 137
Graham, R. N. – see DeWitt
Grandy, R. E.
– (ed.), *Theories and Observation in Science*, Englewood Cliffs, N. J., 1973 – 196
Graves, J. C.
– *The Conceptual Foundations of Contemporary Relativity Theory*, Cambridge, Mass., 1971 – 97
Groenewold, H. J. – 151
Grünbaum, A. – 53-55, 74, 76, 124
– (A) Logical and Philosophical Foundations of the Special Theory of Relativity, in: Danto, Morgenbesser 399-434 – 90
– (B) *Philosophical Problems of Space and Time* (1963), Dordrecht, 1974 – 54, 55, 66, 76, 87, 89, 90, 97, 119, 122, 124, 125
– (C) Popper on Irreversibility, in: Bunge G 316-331 – 125
– (D) *Geometry and Chronometry in Philosophical Perspective*, Minneapolis, 1968 – 40, 54, 74, 76, 87, 90
– (E) Popper's Views on the Arrow of Time, in: Schilpp B, 775-797 – 119, 125, 126
Guldberg, C. M. – 149
Gutting, G.
– Einstein's Discovery of Special Relativity, in: *Phil. Sci.* 39, 1972, 51-68 – 89
Hamilton, W. R. – 135, 180, 189, 190, 192, 202, 204, 205, 209, 215

229

Hanson, N. R.
- (A) The Copenhagen Interpretation of Quantum Theory, in: Danto, Morgenbesser 450-470 – 135, 150, 177, 182
- (B) *Patterns of Discovery*, Cambridge, 1972 – 9, 182
- (C) *Constellations and Conjectures*, Dordrecht, 1973 – 81
- (D) *The Concept of the Positron*, Cambridge, 1963 – 182
Harré, R.
- *Matter and Method*, London, 1964 – 206
Harris, E. E.
- (A) *The Foundations of Metaphysics in Science*, London, 1965 – 213, 214
- (B) *Hypothesis and Perception*, London, 1970 – 9, 10
Heaviside, O. – 142
Hegel, G. W. F. – 1
Heilbron, J. L., Kuhn, T. S.
- The Genesis of the Bohr Atom, in: *Hist. St. Phys.* 1, 1969, 211-290 – 196
Heisenberg, W. – 108, 129, 142, 143, 145, 152, 154, 164, 168, 169, 177, 182, 185 193, 207
- (A) *Die physikalische Prinzipien der Quanten Theorie*, Leipzig, 1930 – 142, 193
- (B) *Das Naturbild der heutigen Physik* (1955), Reinbek, 1962 – 151
- (C) *Physik und Philosophie* (1958), Frankfurt, 1970 – 150, 176, 182, 206, 220, 221
Heitler, W.
- The Departure from Classical Thought in Modern Physics, in: Schilpp A 179-198 – 152, 193
Helmholtz, H. von – 102, 110
- (A) *Die Tatsachen in der Wahrnehmung* (1879); *Zählen und Messen, Erkenntnistheoretisch Betrachtet* (1887), Darmstadt, 1959 – 68
- (B) *Ueber die Erhaltung der Kraft* (1847), Leipzig, 1889 – 107
Hempel, C. G.
- (A) *Fundamentals of Concept Formation in Empirical Science* (1952), Chicago, 1969 – 8, 59, 60, 61, 66, 68-70, 76
- (B) *Aspects of Scientific Explanation*, New York, 1965 – 8, 66, 152, 155, 158 159, 167
- (C) *Philosophy of Natural Science*, Englewood Cliffs, 1966 – 8, 66
- (D) Operationism, Observation, and Scientific Terms, in: Danto, Morgenbesser 101-120 – 66
Henkin, L., Suppes, P., Tarski, A.
- (eds.), *The Axiomatic Method*, Amsterdam, 1959 – 2, 54, 66, 151
Heraclites – 203
Hertz, H. – 110
- *The Principles of Mechanics* (1894), New York, 1956 – 110
Hesse, M.
- (A) *The Structure of Scientific Inference*, London, 1974 – 27, 28, 81, 89, 155, 158
- (B) *Forces and Fields* (1961), Totowa, N. J., 1965 – 88, 110
Hilbert, D. – 43, 45, 46, 169-175, 178-180, 182, 185, 186, 189, 191, 192, 210, 222
Hirosige, S.
- The Ether Problem, the Mechanistic World View, and the Origins of the Theory of Relativity, in: *Hist. St. Phys. Sci.* 7, 1976, 3-82 – 88
Hoenen, P. – 88
- *Philosophie der Anorganische Natuur*, Antwerpen, 1938 – 88

Holton, G. – 23-28, 93
 – (A) *Thematic Origins of Scientific Thought*, Cambridge, Mass., 1973 – 10, 23, 24, 28, 87, 89, 94, 177
 – (B) *The Scientific Imagination*, Cambridge, 1978 – 10, 23
Hooker, C. A.
 – The Nature of Quantum Mechanical Reality, in: Colodny 67-302 – 150, 177, 185
Houtappel, R. M. F., Dam, H. van, Wigner, E. P.
 – The Conceptual Basis and Use of the Geometric Invariance Principles, in: *Rev. Mod. Phys.* 37, 1965, 595-632 – 98, 176
Hume, D. – 9
 – *An Enquiry Concerning Human Understanding* (1748); *A Treatise of Human Nature* (1739), La Salle, Ill., 1971 – 9
Huygens, C. – 6, 97, 111, 144-146, 221
Jaki, S. L.
 – *The Relevance of Physics*, Chicago, 1966 – 20, 206, 220
Jammer, M.
 – (A) *Concepts of Space* (1954), New York, 1960 – 47, 50, 95, 97, 108, 141
 – (B) *concepts of Force* (1957), New York, 1962 – 95, 97, 107, 110, 112, 220
 – (C) *Concepts of Mass* (1961), New York, 1964 – 3, 68, 95, 220
 – (D) *The Conceptual Development of Quantum Mechanics*, New York, 1966 – 115, 123, 135, 142, 146, 150-152, 169, 177, 180, 182, 198, 206, 211, 213, 214, 216
 – (E) *The Philosophy of Quantum Mechanics*, New York, 1974 – 13, 145, 150, 152, 158, 177, 184, 185
Jauch, J. M. – 191
 – *Foundations of Quantum Mechanics*, Reading, Mass., 1968 – 13, 46, 146, 150, 155, 172, 173, 177, 178, 181, 182, 191, 207, 211, 214, 220
Jeffreys, H. – 158
 – *Theory of Probability* (1939), Oxford, 1961 – 158
Jordan, P. – 169
Joule, J. P. – 102
Kaempffer, F. A.
 – *Concepts in Quantum Mechanics*, New York, 1965 – 122, 181, 182, 188
Kalsbeek, L.
 – *Contours of a Christian Philosophy* (1970), Toronto, 1975, – 4
Kant, I. – 18, 20, 110, 204
 – *Kritik der reinen Vernuft* (1781 (A), 1787 (B)), Frankfurt, 1974 – 20, 124
Kelvin, W. – 71, 72, 102, 111
Kepler, J. – 27, 81-83, 110
Keynes, J. M. –158
 – *A Treatise on Probability* (1921), New York, 1962 – 158
Khinchin, A. I.
 – *Mathematical Foundations of Statistical Mechanics*, New York, 1949 – 165
Kirchhoff, G. R. – 110, 115, 144
Kittel, C. – 120
 – *Thermal Physics*, New York, 1969 – 120, 162, 163, 166
Klein, M. J.
 – (A) Einstein and the Wave-Particle Duality, in: *The Natural Philosopher* 3, 1964, 1-49 – 146, 150, 211
 – (B) The First Phase of the Bohr-Einstein Dialogue, in: *Hist. St. Phys. Sci.* 2, 1970, 1-39 – 145, 150
Klein, O. – 137

232

Pfleegor, R. J., Mandel, L.
- Interference of Independent Photon Beams, in: *Phys. Rev.* 159, 1967, 1084-1088 – 147
Pippard, A. B. – 127
- *Elements of Classical Thermodynamics*, Cambridge, 1960 – 127
Planck, M. – 78, 108, 115, 123, 141, 142, 196, 220, 222
Plato – 82
Podolsky, B. – 185, 186 – see also Einstein
Poincaré, H. – 20, 54, 87, 93, 94
- (A) *The Value of Science* (1905), New York, 1958 – 74, 94
- (B) *Science and Hypothesis* (1906), New York, 1952 – 37, 47, 65, 94, 110, 158
Popma, K. J.
- *Nadenken over de Tijd*, Amsterdam, 1965 – 35
Popper, K. R. – 9, 30, 159, 160
- (A) *The Logic of Scientific Discovery* (1934), New York, 1968 – 8, 9, 20, 30, 54, 74, 124, 155, 158, 159, 162, 167
- (B) *Conjectures and Refutations* (1962), New York, 1969 – 8, 220
- (C) Quantum Theory without "The Observer", in: Bunge F 7-44 – 151, 155, 158-160, 165, 192, 196, 198
- (D) Normal Science and its Dangers, in: Lakatos, Musgrave 51-58 – 25
- (E) *Objective Knowledge* (1972), Oxford, 1974 – 9, 29
- (F) Autobiography; Replies to my Critics, in: Schilpp B 3-181, 961-1197 – 1, 124, 125, 160
Price, W. C., Chissick, S. S.
- *The Uncertainty Principle*, London, 1977 – 145
Priestley, J. – 149
Ptolemy – 81, 82
Pythagoras – 6, 26, 41, 82
Ratoosh, P. – see Churchman
Rayleigh, J. W. – 115
Redlich, O.
- Fundamental Thermodynamics since Carathéodory, in: *Rev. Mod. Phys.* 40, 1968, 556-563 – 70, 100, 102
Reichenbach, H. – 77, 118, 119, 123-126, 128, 159
- (A) *The Philosophy of Space and Time* (1928), New York, 1957 – 63, 75-77, 81, 87, 90, 91, 97
- (B) *Philosophic Foundations of Quantum Mechanics* (1944), Berkeley, 1965 – 146
- (C) *Direction of Time* (1956), Berkeley, 1971 – 90, 118, 119, 124-126, 128, 130, 149, 156, 159, 203
- (D) *The Rise of Scientific Philosophy* (1951), Berkeley, 1968 – 145, 206
Riemann, B. – 54
Rosen, N. – 185, 186 – see also Einstein
Russell, B. – 34, 40, 50
- (A) *Einführung in die Mathematische Philosophie* (1919), Darmstadt, no date– 34, 35, 42
- (B) *The Analysis of Matter* (1927), London, 1959 – 9, 97, 141
- (C) *History of Western Philosophy* (1946), London, 1967 – 9, 11, 203
Rutherford, E. – 154, 195
Schaffner, K. F.
- *Nineteenth-Century Aether Theories*, Oxford, 1972 – 88
Scheibe, E.
- Bibliographie zu Grundlagenfragen der Quantenmechanik, in: *Phil. Nat.* 10, 1968, 249-290 – 150

235

Schilpp, P. A.
- (A) (ed.), *Albert Einstein, Philosopher-Scientist* (1949), New York, 1959
- (B) (ed.), *The Philosophy of Karl Popper*, La Salle, Ill., 1974
Schrödinger, E. – 137, 138, 140, 141, 150, 169, 177, 181
Wat ist ein Naturgesetz?, München, 1962 – 119
Scott, W. L.
- *The Conflict between Atomism and Conservation Theory*, Amsterdam, 1970 – 107
Segal, I. E. – 2
Shankland, R. S.
- Conversations with Einstein, in: *Am. J. Phys.* 31, 1963, 47-57; 41, 1973, 895-901 – 88
Sklar, L.
- *Space, Time, and Space-time*, Berkeley, 1974 – 88, 97
Smoluchowski, M. von – 197
Sommerfeld, A. J. W. – 195
Spencer, H. – 110
Spinoza, B. de – 33
Stafleu, M. D.
- (A) Quantumfysica en Wijsbegeerte der Wetsidee, in: *Phil. Ref.* 31, 1966, 126-156
- (B) Individualiteit in de Fysica, in: D. M. Bakker et al., *Reflexies*, Amsterdam, 1968, 288-305
- (C) Analysis of Time in Modern Physics, in: *Phil. Ref.* 35, 1970, 1-24, 119-131
- (D) Metric and Measurement in Physics, in *Phil. Ref.* 37, 1972, 42-57
- (E) *Foundations of Physics, A Christian View*, Toronto, forthcoming
- (F) The Isolation of a Field of Science, *Phil. Ref.*, 44, 1979, 1-15 – 22
- (G) The Mathematical and the Technical Opening up of a Field of Science, *Phil. Ref.* 43, 1978, 18-37 – 22, 71
- (H) The Opening up of a Field of Science by Abstraction and Synthesis, *Phil. Ref.*, 45, 1980, 47-76 – 22, 111
Stefan, J. – 115
Stegmüller, W.
- *Probleme und Resultate der Wissenschaftstheorie und Analytischen Philosophie, Band II: Theorie und Erfahrung*, Berlin, 1970 – 8, 59, 60, 66, 70, 74, 76
Stevens, S. S.
- (A) On the Theory of Scales of Measurement, in: Danto, Morgenbesser 141-149 – 61
- (B) Measurement, Psychophysics and Utility, in: Churchman, Ratoosh 18-63 61, 69
Strauss, D. F. M.
- Number Concept and Number Idea, in: *Phil. Ref.* 35, 1970, 156-177; 36, 1971, 13-42 – 40
Suppes, P. – see also Henkin
- (A) *Introduction to Logic*, New York, 1957 – 36, 60, 62, 69, 101, 110, 155, 156
- (B) Some Open Problems in the Philosophy of Space and Time, in: *Synthese* 24, 1972, 298-316 – 49
Swenson, L. S.
- *The Ethereal Aether*, Austin, 1972 – 88, 89
Tarski, A. – see Henkin
Taylor, B. – 179
Thales – 11
Thomas Aquinas – 88